Ernst Schering Research Foundation Workshop 53
Opportunities and Challenges
of the Therapies Targeting CNS Regeneration

Ernst Schering Research Foundation
Workshop 53

Opportunities and Challenges of the Therapies Targeting CNS Regeneration

H.D. Perez, B. Mitrovic, A. Baron Van Evercooren
Editors

With 19 Figures

Springer

Series Editors: G. Stock and M. Lessl

Library of Congress Control Number: 2004117277

ISSN 0947-6075

ISBN-10 3-540-24082-9 Springer Berlin Heidelberg New York
ISBN-13 978-3-540-24082-2 Springer Berlin Heidelberg New York

Springer is a part of Springer Science+Business Media
springeronline.com

Editor: Dr. Ute Heilmann, Heidelberg
Desk Editor: Wilma McHugh, Heidelberg
Production Editor: Michael Hübert, Leipzig
Cover design: design & production, Heidelberg
Typesetting and production: LE-TEX Jelonek, Schmidt & Vöckler GbR, Leipzig
21/3150/YL – 5 4 3 2 1 0 Printed on acid-free paper

Preface

The therapeutic options for the treatment of multiple sclerosis (MS) and other neurodegenerative and traumatic diseases such as spinal cord injury, Alzheimer's, Parkinson's disease, etc., have undergone enormous progress over recent years. Despite these encouraging developments, available therapies are only partially effective, and the ultimate goal is still far from being attained. Improved understanding of the cellular and molecular mechanisms of the pathogenesis of neurodegeneration and demyelination has led to a variety of new therapeutic targets and approaches.

In addition to modulation of the inflammatory process (MS) and classical neuroprotection (stroke, AD), therapeutic approaches focussing on active remyelinization and neuronal regeneration have become increasingly important. Based on current concepts, this book summarizes new therapeutic approaches.

Although it was once thought that the central nervous system (CNS) of mammals was incapable of substantial recovery from injury, it is now clear that the adult CNS remains responsive to various substances that can promote cell survival and stimulate axonal growth. Among these substances are growth factors, including the neurotrophins and cytokines. Stem cell therapies for the induction of remyelinization and neuroregeneration are reviewed. The potential role of a protective immunity in the induction of remyelination and neuroregeneration is also discussed. Different gene therapy approaches for the treatment of MS and other neurodegenerative diseases such as Alzheimer's disease and spinal cord injury, etc., are also summarized. A major driving force is the understanding of the common regenerative processes that are important for the functional recovery of CNS, regardless of the initial cause of the damage.

Therefore, the molecular and cellular events leading to the demyelination, axonal loss, and neuronal death in different neurodegenerative diseases are also reviewed.

H.D. Perez,
B. Mitrovic

Contents

List of Editors and Contributors

Editors

Perez, H.D.
Berlex Biosciences, Richmond, CA 94806, USA
(e-mail: Daniel_Perez@Berlex.com)

Mitrovic, B.
Berlex Biosciences, Richmond, CA 94806, USA
(e-mail: Branka_Mitrovic@Berlex.com)

Baron Van Evercooren, A.
INSERM, Unit 546, CHU Pitié-Salpétrière, 105 Boulevard de l'Hôpital,
75634 Paris, France
(e-mail: baron@ccr.jussieu.fr)

Contributors

Arnold, D.L.
Montreal Neurological Institute, McGill University, 3801 University Street,
Montreal, Quebec H3A 2B4, Canada
(e-mail: doug@mrs.mmi.mcgill.ca)

Bacigaluppi, M.
Neuroimmunology Unit, Department of Neurology,
San Raffaele Scientific Institute-DIBIT, Via Olgettina, 58, 20132 Milan, Italy

Bucello, S.
Neuroimmunology Unit, Department of Neurology,
San Raffaele Scientific Institute-DIBIT, Via Olgettina, 58, 20132 Milan, Italy

Butti, E.
Neuroimmunology Unit, Department of Neurology,
San Raffaele Scientific Institute-DIBIT, Via Olgettina, 58, 20132 Milan, Italy

Chen, J.
Montreal Neurological Institute, McGill University, 3801 University Street,
Montreal, Quebec H3A 2B4, Canada

Decker, L.
INSERM U368, Ecole Normale Supérieure, 46 rue d'Ulm,
75230 Paris, Cedex 05, France
(e-mail: decker@wotan.ens.fr)

Deleidi, M.
Neuroimmunology Unit, Department of Neurology,
San Raffaele Scientific Institute-DIBIT, Via Olgettina, 58, 20132 Milan, Italy

Duncan, I.D.
Department of Medical Sciences, University of Wisconsin-Madison,
School of Veterinary Medicine, 2015 Linden Drive, Madison, WC 53705-1102

Federoff, H.J.
Center for Aging and Developmental Biology, University of Rochester,
School of Medicine and Dentistry,
601 Elmwood Avenue, Box 645, Room I-9619, Rochester, NY 14642, USA
(e-mail: Howard_Federoff@URMC.rochester.edu)

Furlan, R.
Neuroimmunology Unit, Department of Neurology,
San Raffaele Scientific Institute-DIBIT, Via Olgettina, 58, 20132 Milan, Italy

Goldman, S.A.
Division of Cell and Gene Therapy, Department of Neurology,
University of Rochester Medical Center,
601 Elmwood Avenue, MRBX, Box 645, Rochester, NY 14642, USA
(e-mail: Steven_Goldman@urmc.rochester.edu)

Hohlfeld, R.
Department of Neuroimmunology, Max-Planck-Institute for Neurobiology, Am Klopferspitz 18A, 82152 Martinsried, Germany
and
Institute of Clinical Neuroimmunology and Department of Neurology, Klinikum Grosshadern, Ludwig-Maximilians University, Marchioninistr. 15, 81366 Munich, Germany
(e-mail: reinhard.hohlfeld@nro.med.uni-muenchen.de)

Imitola, J.
Center for Neurologic Diseases, Brigham and Women's Hospital, Department of Nerology, Harvard Medical School, Boston, MA 02115, USA

Kerschensteiner, M.
Institute of Clinical Neuroimmunology and Department of Neurology, Klinikum Grosshadern, Ludwig-Maximilians University, Marchioninistr. 15, 81366 Munich, Germany
and
Department of Neuroimmunology, Max-Planck-Institute for Neurobiology, Am Klopferspitz 18A, 82152 Martinsried, Germany

Khoury, S.J.
Center for Neurologic Diseases, Brigham and Women's Hospital, Department of Nerology, Harvard Medical School, Boston, MA 02115, USA

Lachapelle, F.
INSERM U546, Laboratoire des Affections de la Myéline et des Canaux Ioniques Musculaires, IFR 70, CHU Pitié-Salpétrièere, 105 Boulevard de l'Hôpital, 75634 Paris, France
(e mail: lachapcl@ccr.jussicu.fr)

Lassmann, H.
Division of Neuroimmunology, Brain Research Institute, University of Vienna, Spitalgasse 4, A-1090 Vienna, Austria
(e-mail: hans.lassmann@univie.ac.at)

Loring, F.
Program in Developmental Regenerative Cell Biology, The Burnham Institute, 10901 North Torrey Pines Road, Room 7261, La Jolla, CA 92037, USA

Maguire-Zeiss, K.A.
Center for Aging and Developmental Biology, University of Rochester,
School of Medicine and Dentistry,
601 Elmwood Avenue, PO Box 645, Room I-9609, Rochester, NY 14642, USA
(e-mail: Kathleen_maguirezeiss@urmc.rochester.edu)

Magy, L.
Laboratoire de Neurology CHRU Dupuytren, 2 avenue Martin Luther King,
87042 Limoges, France
(e-mail: Laurent.magy@unilim.fr)

Martino, G.
Neuroimmunology Unit, Department of Neurology,
San Raffaele Scientific Institute-DIBIT, Via Olgettina, 58, 20132 Milan, Italy

McKercher, S.R.
Program in Degenerative Disease Research, The Burnham Institute,
10901 North Torrey Pines Road, La Jolla, CA 92037, USA

Mueller, F.J.
Program in Developmental Regenerative Cell Biology, The Burnham Institute,
10901 North Torrey Pines Road, Room 7261, La Jolla, CA 92037, USA
(e-mail: franz-josef.mueller@gmx.de)

Nait-Oumesmar, B.
INSERM U546, Laboratoire des Affections de la Myéline
et des Canaux Ioniques Musculaires, IFR 70, CHU Pitié-Salpêtrière,
105 Boulevard de l'Hôpital, 75634 Paris, France
(e-mail: brahim.nait-oumesmar@chups.jussieu.fr)

Picard-Riera, N.
INSERM U546, Laboratoire des Affections de la Myéline
et des Canaux Ioniques Musculaires, IFR 70, CHU Pitié-Salpêtrière,
105 Boulevard de l'Hôpital, 75634 Paris, France
(e-mail: npicard@snv.jussieu.fr)

Pluchino, S.
Neuroimmunology Unit, Department of Neurology,
San Raffaele Scientific Institute-DIBIT, Via Olgettina, 58, 20132 Milan, Italy

Sim, F.J.
Division of Cell and Gene Therapy, Department of Neurology,
University of Rochester Medical Center,
601 Elmwood Avenue, MRBX, Box 645, Rochester, NY 14642, USA

Snyder, E.Y.
Program in Developmental Regenerative Cell Biology, The Burnham Institute,
10901 North Torrey Pines Road, Room 7261, La Jolla, CA 92037, USA
(e-mail: esnyder@burnham.org)

Stadelmann, C.
Institute of Neurology, University Vienna, Schwarzspanierstr. 17,
1090 Vienna, Austria

Tuszynski, M.
Department of Neurosciences, Center for Neural Repair,
University of Californie, San Diego, 9500
Gilman Drive, La Jolla, CA 92093-0626, USA
(e-mail: mtuszynski@ucsd.edu)

Wekerle, H.
Department of Neuroimmunology, Max-Planck-Institute for Neurobiology,
Am Klopferspitz 18A, 82152 Martinsried, Germany

Yip, S.
Division of Neuropathology, Department of Pathology,
Vancouver General Hospital, University of British Columbia, Vacouver, BC,
Canada

Zanotti, L.
Neuroimmunology Unit, Department of Neurology,
San Raffaele Scientific Institute-DIBIT, Via Olgettina, 58, 20132 Milan, Italy

1 New Strategies for CNS Repair

M.H. Tuszynski

1.1 Introduction

A number of neurological disorders lack effective therapies. This is true of many neurodegenerative disorders, including Alzheimer's disease and Parkinson's disease, in which drugs to prevent or reduce disease progression do not exist. Other disorders of the nervous system inflict damage as a result of a single event in time, including stroke, head trauma, and spinal cord injury; in these cases, there is a great need for the development of effective therapies to restore lost function. Finally, there are neurological diseases such as multiple sclerosis in which drugs have been developed to slow disease progression, yet a great need remains to improve or restore functions that continue to slowly decline over time.

The transplantation of replacement cells into the adult nervous system, or the targeted delivery of therapeutic genes to areas of ongoing degeneration, have received considerable attention over the last several years as potential means of enhancing the treatment of central nervous system (CNS) disease. The development of improved vector systems for in vivo gene delivery, the discovery and development of neural stem cells, and the unraveling of basic mechanisms underlying disease progression or secondary injury have opened new possibilities for therapeutic intervention in progressive neurodegenerative disease and nervous system injury. Strategies have evolved for the delivery of potentially neuroprotective molecules, such as neurotrophic factors, and the replacement of cells and tissue lost due to CNS injury and degeneration, and some of these efforts are beginning the transition to clinical trials. This chapter will highlight developments in one of these areas: the therapeutic use of growth factors to reduce cell loss and to enhance axonal regeneration in the context of both neurodegenerative and traumatic disorders.

1.2 Growth Factor Neuroprotection in Neurodegenerative Disease

Neuroprotective strategies are of particular potential interest in neurodegenerative diseases such as Alzheimer's disease (AD) and Parkinson's disease (PD), where neurons of the central nervous system undergo degeneration over protracted time periods of several years. This prolonged period of neurodegeneration presents the opportunity to intervene early in the disease to reduce cell loss and possibly improve function.

The discovery in the mid-1980s that growth factors can prevent the death of degenerating neurons in the adult brain opened new possibilities for the treatment of neurological disease (Hefti 1986; Williams et al. 1986; Kromer et al. 1987). The first growth factor to show efficacy in the adult brain was nerve growth factor (NGF), which demonstrated an ability to prevent the death of virtually 100 % of cholinergic neurons in the basal forebrain after injury (Levi-Montalcini 1987). NGF subsequently was shown to reverse age-related neuronal atrophy in rodents (Fischer et al. 1987), to ameliorate cholinergic degeneration resulting from excessive amyloid levels in the brain (Holtzman et al. 1993), and to improve learning and memory in both aged and lesioned rats (Holtzman et al. 1993; Tuszynski and Gage 1995). Further, it retained the ability to prevent cholinergic neuron death in the adult primate brain (Tuszynski et al. 1990, 1991; Koliatsos et al. 1991).

Notably, cholinergic cell loss is a cardinal component of neuronal dysfunction in AD (Perry et al. 1977), and loss of cholinergic neurons is thought to contribute to cognitive decline, particularly in early and mid stages of AD (Bartus et al. 1982; Mufson et al. 2002). Thus, the discovery that NGF can prevent the loss of cholinergic neurons established a clear rationale for its potential use as a neuroprotective therapy for AD. While a number of other cell systems that are distributed throughout the cortex also degenerate in AD, it is possible that targeting the cholinergic component of cell loss with NGF will be sufficient to have a significant impact on cognitive decline and quality of life. Indeed, most drugs currently approved for the treatment of AD target only these cholinergic neurons, and while only modestly elevating acetylcholine levels in the brain all of these drugs have detectably beneficial effects on cognition (Davis et al. 1992). Thus, it is reasonable to hypothesize that a therapy aimed at more potently stimulating these cells, while also reducing their ongoing degeneration over time, will be beneficial in AD.

To be effective, growth factors must be delivered directly into the brain due to the relatively large size and polarity of these proteins, which prohibits their passage across the blood brain barrier. Moreover, to avoid adverse effects of growth factors that result when the proteins are widely distributed throughout the neuraxis (reviewed in Tuszynski 2002), their central delivery must be restricted to only the targeted neuronal region. Otherwise, broad distribution of growth factors such as NGF causes pain, weight loss, and Schwann cell migration into the CNS as a result of stimulation of growth factor receptors on nontargeted cell systems.

One means of directly delivering therapeutic substances into the CNS directly, and restricting their delivery to targeted regions, is gene therapy. Several studies have been performed using methods of ex vivo NGF gene delivery, demonstrating that such cellularly-delivered NGF was indeed effective in preventing the degeneration of basal forebrain cholinergic neurons and ameliorating behavioral deficits (Rosenberg et al. 1988; Chen and Gage 1995). Subsequent studies in monkeys determined that NGF delivered by genetically modified cells prevents lesion-induced degeneration of medial septal cholinergic neurons (Emerich et al. 1994; Tuszynski et al. 1996a), as reported previously with NGF infusions. Further, NGF delivery in primates by grafts of genetically modified primary fibroblasts reversed the age-related atrophy of cholinergic neurons in the nucleus basalis (Smith et al. 1999) and restored cortical cholinergic innervation to levels of young monkeys (Conner et al. 2001). To determine whether NGF gene delivery was safe and lacked significant toxicity, additional monkeys received grafts of autologous fibroblasts expressing NGF for time periods of 1 year. Over a wide range of injected cell volumes, no signs of toxicity were detected. Specifically, weight loss, pathological responses of Schwann cells, and general indices of pain were not observed in primates. NGF did not leak into the cerebrospinal fluid when genetically modified cells were injected into brain parenchyma, cellular grafts did not form tumors, and cells did not migrate from the injection site (Tuszynski 2002).

Thus, evidence from rodent and primate studies suggested that NGF delivered by genetically modified fibroblasts in a well-targeted, spatially restricted, and intraparenchymal manner to target neurons in the basal forebrain was effective in slowing or preventing the degeneration of cholinergic neurons. Further, this approach was safe over prolonged time periods in primates. These findings provided a basis for translating this therapeutic approach to humans with AD, with the aims of: (1) testing whether NGF delivery to the human brain could prevent cholinergic cell death via whatever mechanism causes neuronal loss in AD, and (2) determining whether ameliorating the cholinergic component of neuronal degeneration in AD would be sufficient to meaningfully reduce disease progression and improve quality of life.

A phase I clinical trial of NGF gene delivery for AD has now been conducted at the University of California, San Diego. Eight patients with early stage AD were enrolled in the study. Patients received increasing doses of NGF- secreting autologous fibroblasts implanted in the region of the nucleus basalis of Meynert. To date, no long-term toxicities associated with NGF delivery have been observed in the clinical trial, with a period of observation between 18 months and 3 years in all patients. Efficacy parameters are currently under study.

The studies described above focused on ex vivo gene therapy rather than in vivo gene therapy, because ex vivo gene delivery was better developed than in vivo gene delivery at the time these studies were initiated. Yet in vivo gene delivery is less complex than ex vivo methods. Animal studies conducted with in vivo gene delivery vectors, including AAV and lentiviral vectors, support the fact that in vivo NGF gene delivery is equally as effective as protein infusion and ex vivo gene delivery in preventing degeneration of cholinergic neurons (Blomer et al. 1998; Klein et al. 2000). Additional safety and toxicity studies have also been conducted using in vivo NGF gene delivery in rodents and primates, yielding supportive evidence of long-term safety. Based on these data, a second human clinical trial of NGF gene delivery to patients with AD has now been initiated, using the AAV vector expressing human NGF to genetically modify cells within the region of the nucleus basalis.

To optimize the safety of gene therapy, it would be desirable to have the ability to regulate the expression of therapeutic genes after gene delivery. Systems such as the tetracycline-, rapamycin- or ecdysone-regulated system are potential candidates, but need further development before entering clinical trials due to persistent practical limitations with each system.

1.3 Growth Factor Stimulation of Axonal Regeneration After Spinal Cord Injury

In contrast to experimental therapies for neurodegenerative diseases that aim to prevent or slow neuronal degeneration, experimental therapies for spinal cord injury attempt to augment the regeneration and growth of injured axons. Inhibitors of axonal growth are present in adult CNS myelin (Schwab and Bartholdi 1996) and in the inhibitory extracellular matrix that surrounds the site of spinal cord injury (Fitch et al. 1999; Morgenstern et al. 2002), contributing to the failure of axonal regeneration in the adult spinal cord. Further, cystic degeneration at the injury site requires replacement of a suitable axonal growth substrate to support axonal growth through the lesion cavity.

Neurotrophic factors have been investigated as a means of stimulating axonal growth after SCI to counteract the inhibitory influence of factors that restrict axon growth. By placing cells into the lesion site that have been genetically modified to produce and secrete growth factors, one can both: (1) provide a growth factor locally at the site of injury to stimulate axonal regeneration, and (2) reconstitute a cellular matrix within the lesion cavity over which axons might extend. Fibroblasts, Schwann cells, neural stem cells, and bone marrow stromal cells have all been genetically modified to express and secrete various neurotrophic factors, including NGF, brain-derived neurotrophic factor (BDNF), neurotrophion-3 (NT-3), neurotrophin-4/5 (NT-4/5), glial cell line-derived neurotrophic factor (GDNF), leukemia inhibitory factor (LIF), and the cytokine growth factor interleukin-6 (IL-6) (Tuszynski et al. 1994, 1996b, 2003; Grill et al. 1997; Jakeman et al. 1998; McTigue et al. 1998; Blesch et al. 1999, 2003; Liu et al. 1999; Weidner et al. 1999; Lu et al. 2001, 2003, 2004; Lacroix et al. 2002). When grafted to sites of SCI in adult rats, many of these growth factors have acted as a supportive growth substrate for injured axons and have induced robust axonal growth. In some cases, functional recovery has been observed, although the magnitude of this recovery has been modest (Grill et al. 1997; Jakeman et al. 1998; Liu et al. 1999; Tuszynski et al. 2003). Nonetheless, these experiments have clearly established the principle that growth factor gene delivery to sites of SCI can be an important component of a combinatorial strategy to alter a site of CNS injury and convert a nonpermissive to a permissive milieu for axonal regeneration.

Although axonal growth is robust into many neurotrophin-expressing cellular grafts, axonal growth distal to the lesion site is not observed. Yet axonal growth distal to the lesion site will be a clear requisite for achieving meaningful and extensive motor improvement. Recently, we found that the provision of a gradient of growth factor beyond a spinal cord injury and graft site was required to achieve meaningful bridging of host axons (Lu et al. 2004). Further optimization of this approach to stimulate axonal growth, in combination with methods such as cyclic nucleotide administration to injured neurons (Lu et al. 2004; Pearse et al. 2004), may lead to the optimization of strategies for enhancing CNS repair. Continued careful and objective studies of this sort in the future may eventually lead to a clear rationale for human clinical trials in SCI.

1.4 Conclusion

Advances in the neurosciences have led to new therapeutic possibilities for diseases that have been untreatable. While the above discussion has considered the specific instances of AD and spinal cord injury, promising parallel findings have been reported in animal models of PD, amyotrophic lateral sclerosis, multiple sclerosis, Huntington's disease and others. The next two decades are likely to represent a golden era of molecular medicine that will change the landscape of neurological diagnosis and therapy.

Acknowledgements. This work was supported by the NIH, Veterans Affairs, the Shiley Family Foundation and the Institute for Research on Aging.

References

Bartus R, Dean RL, Beer C, Lippa AS (1982) The cholinergic hypothesis of geriatric memory dysfunction. Science 217:408–417

Blesch A, Tuszynski MH (2003) Cellular GDNF delivery promotes growth of motor and dorsal column sensory axons after partial and complete spinal cord transections and induces remyelination. J Comp Neurol 467:403–417

Blesch A, Uy HS, Grill RJ, Cheng JG, Patterson PH, Tuszynski MH (1999) Leukemia Inhibitory Factor augments neurotrophin expression and corticospinal axon growth after adult CNS injury. J Neurosci 19:3556–3566

Blomer U, Kafri T, Randolph-Moore L, Verma IM, Gage FH (1998) Bcl-xL protects adult septal cholinergic neurons from axotomized cell death. Proc Natl Acad Sci 95:2603–2608

Chen KS, Gage FH (1995) Somatic gene transfer of NGF to the aged brain: Behavioral and morphological amelioration. J Neurosci 15:2819–2825

Conner JM, Darracq MA, Roberts J, Tuszynski MH (2001) Non- tropic actions of neurotrophins: Subcortical NGF gene delivery reverses age- related degeneration of primate cortical cholinergic innervation. Proc Nat Acad Sci 98:1941–1946

Davis KL, Thal LJ, Gamzu ER, Davis CS, Woolson RF, Gracon SI, Drachman DA, Schneider LS, Whitehouse PJ, Hoover TM (1992) A double-blind, placebo-controlled multicenter study of tacrine for Alzheimer's disease. The Tacrine Collaborative Study Group. New Engl J Med 327:1253–1259

Emerich DW, Winn S, Harper J, Hammang JP, Baetge EE, Kordower JH (1994) Implants of polymer-encapsulated human NGF-secreting cells in the nonhuman primate: Rescue and sprouting of degenerating cholinergic basal forebrain neurons. J Comp Neurol 349:148–164

Fischer W, Wictorin K, Bjorklund A, Williams LR, Varon S, Gage FH (1987) Amelioration of cholinergic neuron atrophy and spatial memory impairment in aged rats by nerve growth factor. Nature 329:65–68

Fitch S, Silver J (1999) Beyond the glial scar. In: Tuszynski MH, Kordower J (eds) CNS regeneration: Basic science and clinical advances. Academic Press, San Diego, pp 55–88

Grill R, Murai K, Blesch A, Gage FH, Tuszynski MH (1997) Cellular delivery of neurotrophin-3 promotes corticospinal axonal growth and partial functional recovery after spinal cord injury. J Neurosci 17:5560–5572

Hefti F (1986) Nerve growth factor (NGF) promotes survival of septal cholinergic neurons after fimbrial transection. J Neurosci 6:2155–2162

Holtzman DM, Li Y, Chen K, Gage FH, Epstein CJ, Mobley WC (1993) Nerve growth factor reverses neuronal atrophy in a Down syndrome model of age-related neurodegeneration. Neurology 43:2668–2673

Jakeman LB, Wei P, Guan Z, Stokes BT (1998) Brain-derived neurotrophic factor stimulates hindlimb stepping and sprouting of cholinergic fibers after spinal cord injury. Exp Neurol 154:170–184

Klein RL, Hirko AC, Meyers CA, Grimes JR, Muzyczka N, Meyer EM (2000) NGF gene transfer to intrinsic basal forebrain neurons increases cholinergic cell size and protects from age-related, spatial memory deficits in middle-aged rats. Brain Res 875:144–151

Koliatsos VE, Clatterbuck RE, Nauta HJ, Knusel B, Burton LE, Hefti FF, Mobley WC, Price DL (1991) Human nerve growth factor prevents degeneration of basal forebrain cholinergic neurons in primates. Ann Neurol 30:831–840

Kromer LF (1987) Nerve growth factor treatment after brain injury prevents neuronal death. Science 235:214–216

Lacroix S, Chang L, Rose-John S, Tuszynski MH (2002) Delivery of hyper-interleukin-6 to the injured spinal cord increases neutrophil and macrophage infiltration and inhibits axonal growth. J Comp Neurol 454:213–228

Levi-Montalcini R (1987) The nerve growth factor 35 years later. Science 237:1154–1162

Liu Y, Kim D, Himes BT, Chow SY, Schallert T, Murray M, Tessler A, Fischer I (1999) Transplants of fibroblasts genetically modified to express BDNF promote regeneration of adult rat rubrospinal axons and recovery of forelimb function. J Neurosci 19:4370–4387

Lu P, Blesch A, Tuszynski MH (2001) Neurotrophism Without Neurotropism: BDNF promotes survival but not growth of lesioned corticospinal neurons. J Comp Neurol 436:456–470

Lu P, Jones LL, Snyder EY, Tuszynski MH (2003) Neural stem cells constitutively secrete neurotrophic factors and promote extensive host axonal growth after spinal cord injury. Exp Neurol 181:115–129

Lu P, Yang H, Jones LL, Filbin MT, Tuszynski MH (2004) Combinatorial therapy with neurotrophins and cAMP promotes axonal regeneration beyond sites of spinal cord injury. J Neurosci 24:6402–6409

McTigue DM, Horner PJ, Stokes BT, Gage FH (1998) Neurotrophin- 3 and brain-derived neurotrophic factor induce oligodendrocyte proliferation and myelination of regenerating axons in the contused adult rat spinal cord. J Neurosci 18:5354–5365

Morgenstern DA, Asher RA, Fawcett JW (2002) Chondroitin sulphate proteoglycans in the CNS injury response. Prog Brain Res 137:313–332

Mufson EJ, Ma SY, Dills J, Cochran EJ, Leurgans S, Wuu J, Bennett DA, Jaffar S, Gilmor ML, Levey AI, Kordower JH (2002) Loss of basal forebrain P75(NTR) immunoreactivity in subjects with mild cognitive impairment and Alzheimer's disease. J Comp Neurol 443:136–153

Pearse DD, Pereira FC, Marcillo AE, Bates ML, Berrocal YA, Filbin MT, Bunge MB (2004) cAMP and Schwann cells promote axonal growth and functional recovery after spinal cord injury. Nat Med 10:610–616

Perry EK, Perry RH, Blessed G, Tomlinson BE (1977) Necropsy evidence of central cholinergic deficits in senile dementia. Lancet 2:143

Rosenberg MB, Friedmann T, Robertson RC, Tuszynski M, Wolff JA, Breakefield XO, Gage FH (1988) Grafting genetically modified cells to the damaged brain: restorative effects of NGF expression. Science 242:1575–1578

Schwab ME, Bartholdi D (1996) Degeneration and regeneration of axons in the lesioned spinal cord. Physiol Rev 76:319–370

Smith DE, Roberts J, Gage FH, Tuszynski MH (1999) Age-associated neuronal atrophy occurs in the primate brain and is reversible by growth factor gene therapy. Proc Nat Acad Sci 96:10893–10898

Tuszynski MH (2002) Gene therapy for neurodegenerative disorders. Lancet Neurol 2002:51–57

Tuszynski MH, Gage FH (1995) Bridging grafts and transient NGF infusions promote long-term CNS neuronal rescue and partial functional recovery. Proc Nat Acad Sci 92:4621–4625

Tuszynski MH, Sang H, Yoshida K, Gage FH (1991) Recombinant human nerve growth factor infusions prevent cholinergic neuronal degeneration in the adult primate brain. Ann Neurol 30:625–636

Tuszynski MH, U HS, Amaral DG, Gage FH (1990) Nerve growth factor infusion in primate brain reduces lesion-induced cholinergic neuronal degeneration. J Neurosci 10:3604–3614

Tuszynski MH, Roberts J, Senut MC, U H-S, Gage FH (1996a) Gene therapy in the adult primate brain: intraparenchymal grafts of cells genetically modified to produce nerve growth factor prevent cholinergic neuronal degeneration. Gene Therapy 3:305–314

Tuszynski MH, Peterson DA, Ray J, Baird A, Nakahara Y, Gage FH (1994) Fibroblasts genetically modified to produce nerve growth factor induce robust neuritic ingrowth after grafting to the spinal cord. Exp Neurol 126:1–14

Tuszynski MH, Gabriel K, Gage FH, Suhr S, Meyer S, Rosetti A (1996b) Nerve growth factor delivery by gene transfer induces differential outgrowth of sensory, motor and noradrenergic neurites after adult spinal cord injury. Exp Neurol 137:157–173

Tuszynski MH, Grill R, Jones LL, Brant A, Blesch A, Low K, Lacroix S, Lu P (2003) NT-3 gene delivery elicits growth of chronically injured corticospinal axons and modestly improves functional deficits after chronic scar resection. Exp Neurol 181:47–56

Weidner N, Blesch A, Grill R, Tuszynski, MH (1999) NGF- Hypersecreting Schwann cell grafts augment and guide spinal cord axonal growth, and remyelinate CNS axons in a phenotypically appropriate manner that correlates with expression of L1. J Comp Neurol 413:495–506

2 Heterogeneity of Multiple Sclerosis: Implications for Therapy Targeting Regeneration

H. Lassmann

2.1 Introduction

Multiple sclerosis (MS) is a chronic inflammatory disease, which leads to widespread, plaque-like demyelination in the central nervous system (Charcot 1868; Lassmann 1998). The hallmark of structural changes is primary demyelination with relative axonal preservation in focal plaques of the white matter. Axonal preservation, however, is relative in relation to the complete destruction of myelin. This implies that in all MS lesions there is axonal destruction and loss, although its extent varies between different patients and between different lesions within a single patient. The process of demyelination in MS lesions is counterbalanced

Table 1. Heterogeneity of multiple sclerosis lesions

Type	Stage dependent	Inter-individual	Mechanism
Immunopathology			
CD 8-dominated inflammation	+	+/–	Basic inflammatory reaction similar in all cases
Macrophage activation	+++	+/–	Macrophage activation in all active lesions
Antibody and complement	–	+++	Immunogenetics?
Hypoxia-like tissue injury	–	+++	Immunogenetics? Extent of damage through oxygen and nitrogen radicals?
Extent of tissue damage	+	++	Genetic susceptibility of target tissue? Apo E? CNTF? others?
Structural			
Oligodendrocyte loss	+	++	Immunological mechanisms of tissue injury?
Axonal injury	++	++	Recurrent demyelination? Ongoing axonal destruction in established plaques? Suscepti-bility of target tissue?
Remyelination	+	++	Recurrent demyelination? Immunological mechanisms of tissue injury? Genetic susceptibility of target tissue?

by spontaneous remyelination; its extent too is variable between lesions and patients.

Besides focal white matter plaques there is also a diffuse damage in the so called normal white matter as well as in the cerebral cortex. These diffuse changes are sparse in patients with acute or relapsing disease, but become very prominent in the progressive stage of the disease. Also at this stage the destructive process is associated with a diffuse inflammatory process, which is not restricted to plaques in the white matter, but affects the global brain and meninges. This inflammatory process appears to be compartmentalized in the brain, reflecting chronic persistent inflammatory infiltrates in the meninges and perivascular connective tissue spaces, and it is no more driven by new waves of inflammation, entering the tissue from the circulation. Thus, at this stage there is no major blood brain barrier damage in spite of the persistence of the inflammatory process. Furthermore, the structural and immunopathological features of the diffuse brain injury in progressive MS are different from those seen in focal white matter plaques, which are the hallmark of acute and relapsing MS.

Thus, MS is a disease with a complex pathogenesis, showing profound heterogeneity of the lesions, depending both upon the stage of the disease and the type of disease, which appears to differ between individual patients or patient subgroups (Table 1).

2.2 Heterogeneity in the Mechanisms of Demyelination and Tissue Injury

Neurobiological studies have shown that brain inflammation can lead to demyelination and tissue injury, which can be accomplished by various different means. This involves the direct attack of cytotoxic T-cells against oligodendrocytes or axons (Neumann et al. 2002), antibody-mediated tissue damage through activation of complement or the interaction with activated macrophages (Linington et al. 1988; Archelos and Hartung 2000) as well as the direct toxic action of a variety of different macrophage products, such as cytokines (Probert et al. 2000), proteases (Cuzner and Opdenakker 1999), reactive oxygen (Smith et al. 1999) and nitrogen species (Smith et al. 2001) or excitotoxins (Smith et al. 2000; Pitt et al. 2000). Neuropathological studies in MS have shown that most of the molecules which are involved in these processes can be present in actively demyelinating lesions. However, by analyzing a large series of active MS lesions it became clear that the expression of these molecules vary between different patient populations (Lucchinetti et al. 2000). In addition, the extent and severity of tissue damage may also depend upon genetic modifications of the target tissue, which may render the brain of individual patients more susceptible to immune-mediated damaged compared to that of others (Fazekas et al. 2000; Giess et al. 2002; Linker et al. 2002). On this basis, four different patterns of demyelination have recently been identified, which were similar in different active lesions of the same patient, but differed between patients (Lucchinetti et al. 2000) and were defined as tissue injury mediated by T-cells and activated macrophages (pattern 1), lesions with prominent involvement of antibodies and activated complement (pattern 2; Storch et al. 1998), hypoxia-like tissue injury (pattern 3; Aboul Enein et al. 2003) and lesions with extensive oligodendrocyte destruction in the periplaque white matter at the lesion borders (pattern 4; Lucchinetti et al. 2000). While patterns 1 and 2 in many respects are mimicked in different models of autoimmune encephalomyelitis, the hypoxia-like tissue injury of pattern 3 has so far not been reproduced in any experimental model of inflammatory demyelination. However, extensive oligodendrocyte damage resembling pattern 4 is found in animals with autoimmune encephalitis with a coincident deletion of the gene for the ciliary neurotrophic factor (Linker et al. 2002).

In the global MS population these different patterns do not segregate with specific disease manifestations, such as relapsing or progressive MS. They do, however, segregate clearly in the extreme variants of the disease. All patients with Devic's type of neuromyelitis optica show profound antibody-mediated tissue injury (Lucchinett et al. 2002) and this is also reflected in experimental autoimmune encephalomyelitis. Also in this model, rat strains with very severe antibody responses show a dominant involvement of optic nerves and spinal cord (Storch et al. 1998). On the contrary, a concentric pattern of demyelination, as found in Balo's disease, is always associated with a hypoxia-like tissue injury. Preliminary data suggest that the different immunopathological patterns of demyelination may be reflected in different morphologies and patterns of lesion development in magnetic resonance imaging, but these findings have to be confirmed in larger series of patients.

This immunopathological heterogeneity accounts for the formation of focal white matter plaques. In patients with progressive MS, a slow burning expansion of preexisting plaques can be seen, which seems mainly to be mediated by activated macrophages and microglia cells, thus resembling the demyelination pattern 1 (Prineas et al. 2001). In addition, however, there is a diffuse damage of the so called normal white matter. This is associated with persistent inflammation and a diffuse microglia activation throughout the whole brain (H. Lassmann, unpublished). Interestingly, in this diffuse damage axonal injury is prominent, while primary demyelination is sparse. The diffuse axonal injury may in part reflect Wallerian degeneration, due to axonal destruction and loss within the plaques (Evangelou et al. 2000a). However, in many cases diffuse axonal injury is present in the "normal white matter", which is so widespread that it cannot be explained on the basis of Wallerian degeneration. It is thus likely that in progressive MS a direct diffuse axonal injury in the whole CNS white matter occurs. The immunological mechanisms behind these process are so far undefined, but they apparently differ from those seen in focal white matter lesions.

All these data show that there is a profound heterogeneity in the immunopathology of MS lesions, which to an extent may reflect the stage and severity of the disease process. To a larger extent, however, these differences are seen between different patients and suggest an interindividual heterogeneity of the disease process.

2.3 Heterogeneity in the Quality of Tissue Injury and Repair

Besides these differences in immunopathology, a profound variability in the quality and extent of tissue injury and repair can be seen between different plaques of a single patient or between different MS patients. These differences involve demyelination and oligodendrocyte injury, remyelination, and axonal damage.

2.3.1 Oligodendrocyte Injury and Remyelination

While demyelination is the hallmark of all MS plaques, the extent of oligodendrocyte destruction and loss is highly variable between different lesions and patients (Lucchinetti et al. 1999). The extent of oligodendrocyte loss within a given lesion in part depends upon the stage of lesion development. Thus, during the stage of active demyelination, oligodendrocytes are in part destroyed together with myelin sheaths, yet the extent of oligodendrocyte destruction is variable. At later stages new oligodendrocytes reappear within the lesions in some patients and these cells are most likely derived from the pool of progenitor cells (Wolswijk 2002). However, similar to immunopathology, the differences in oligodendrocyte loss are most pronounced between different MS patients and much less dramatic between different lesions of a single patient (Lucchinetti et al. 1999). Thus, in some patients the density of oligodendrocytes within the lesions is very high and this is associated with rapid and effective remyelination. In such patients up to 80 % of the focal lesions within the CNS are remyelinated shadow plaques. In others, however, oligodendrocytes are nearly completely lost and remyelination is absent.

To what extent remyelination is associated with differences in clinical course and severity of the disease is so far not established. In general, remyelination is sparse in patients with long-standing progressive MS. However, extensive remyelination can even be found in single patients with a very severe progressive disease course.

Several different factors may determine the extent of remyelination in MS lesions (Franklin 2002). One factor is obviously the availability of oligodendrocytes or their progenitor cells (Wolswijk 2002). In cases where the progenitor cells are simultaneously destroyed with myelin and mature oligodendrocytes in active lesions, remyelination will be sparse. Such destruction of progenitor cells may either be mediated through certain macrophage toxins or by specific antibodies. Interestingly, antibodies against AN-2, a molecule expressed on the surface of progenitor cells, have been found in a subset of MS patients (Niehaus et al. 2000). It will have to be determined in the future, whether in such patients remyelination is absent. However, in many MS plaques oligodendrocyte progenitors are present, which remain in an undifferentiated stage and fail to remyelinate (Chang et al. 2002). It is suggested that in such lesions the failure of remyelination is due to a defect of axons to induce the myelination process (Charles et al. 2002).

2.3.2 Axonal Injury and Loss

Axonal injury is present in MS patients within demyelinated plaques as well as diffusely in the “normal white matter” (Kornek and Lassmann 1999). Within the plaques massive acute axonal injury occurs during the phase of active demyelination (Ferguson et al. 1997; Trapp et al. 1998). In addition, there is a low grade chronic axonal loss and injury in inactive demyelinated lesions (Kornek et al. 2000). The extent of axonal loss within MS lesions is highly variable. It is most pronounced in old, inactive demyelinated plaques, where in average between 50 %–60 % of all axons are lost, while it is relatively mild in remyelinated shadow plaques and in fresh lesions. These differences in axonal loss may be explained by two mechanisms. The ongoing axonal destruction within established demyelinated lesions will accumulate in progressive axonal loss (Kornek et al. 2000). In addition, remyelinated plaques may become targets of recurrent demyelination (Prineas et al. 1993a), each episode leading to profound axonal injury during the phase of active myelin destruction. Thus, the heterogeneity of axonal loss between different plaques of a single patient can largely be explained on this basis. In addition, however, the degree of axonal loss is also highly variable between plaques from different patients, which are exactly matched regarding their stage. Thus, it seems that in some patients axonal injury is more extensive than in others. Whether this depends upon differences in the severity of the inflammatory process or upon a different vulnerability of the target tissue is so far undetermined.

2.4 Genetic Heterogeneity of Multiple Sclerosis

Genetic factors play an important role in the pathogenesis of MS (Compston 2004). However, multiple different genes appear to determine disease susceptibility, disease course, or severity. Although the exact role of individual genes in the pathogenesis of this disease is so far unknown, recent data indicate that different classes of genes may be involved (Compston 2004). They include either genes that regulate the immune response or others, which may determine the susceptibility of the CNS tissue to immune-mediated attack. It is likely that different combinations of genetic influences determine or modify the disease phenotype in different patients, even when the initiating event or the basic mechanisms of disease pathogenesis are the same. This could explain at least to some extent the heterogeneity in pathology described above.

There are some examples which illustrate this concept. The apolipoprotein ε4 genotype has been described to be associated with more severe disease and more destructive lesions in MRI images (Fazekas et al. 2000). Whether this is due to differences in the extent of axonal injury or differences in remyelination has to be determined. In a similar way, a deletion of the gene for ciliary neurotrophic factor is associated with earlier disease onset and increased disease severity (Giess et al. 2002). Experimental studies show that its deletion results in more pronounced oligodendrocyte injury during T-cell mediated brain inflammation (Linker et al. 2002). Comparable effects on brain tissue susceptibility in inflammatory lesions may underlie the observed association between MS and mutations in mitochondrial DNA (Mojon et al. 1999) or the SCA2 gene (Chataway et al. 1999). Other studies suggest that some aspects of disease heterogeneity could be based on differences in HLA associations (Compston 2004). So far, however, detailed studies on the effect of defined genetic polymorphisms or mutations on the pathology of MS lesions are missing.

2.5 Implications for Therapy Targeting Regeneration

The concept of heterogeneity of MS lesions has major implications for therapeutic strategies aimed to stimulate remyelination and repair. First, it implies that such strategies may work in some patients but fail in others. Thus, the question whether transplanted stem or progenitor cells may survive or be rejected may depend upon the mechanism and the specificity of the immune response in a given patient. For instance, in patients with a strong pathogenic antibody response against oligodendrocytes or progenitor cells, the chance for such a transplant to survive and directly remyelinate the tissue is likely to be poor. Secondly, endogenous repair appears to be effective in some patients, but may fail in others. As mentioned above, spontaneous remyelination is pronounced in some patients, while it is absent in others. Obviously it makes no sense to stimulate remyelination in a patient, whose lesions are effectively repaired by endogenous mechanisms. Thirdly, therapeutic strategies to stimulate remyelination and repair may work in some patients, but not in others. As an example, stimulation of oligodendrocyte progenitors to promote remyelination may have little effect in patients in whom the axons are not permissive to become remyelinated. Unfortunately, so far the clinical tools are not established which enable stratification of patients according to the above-listed criteria. Thus, future research has to address a variety of essential points.

Paraclinical tools have to be developed and improved which allow an in vivo imaging of the pathology of MS lesions. In particular, it is impossible so far to unequivocally identify remyelination in MS lesions in magnetic resonance imaging. In addition, more sensitive techniques have to be developed to determine quantitatively the extent of axonal loss in MS lesions. Similarly, it will be important to find immunological and imaging tools that will define the different patterns and mechanisms of demyelination and tissue injury during the patient's life.

Regarding the development of new therapeutic targets for remyelination and repair, it will be essential to validate the role of the respective targets within the MS lesions. Considering the heterogeneity of the disease this, however, will have to be done on a large sample of well-defined patients and exactly staged lesions. This approach is even more important for future studies on gene or protein expression using microarray or proteomics techniques.

Finally, it will be instrumental in the future to study genetic association with the pathological lesion phenotype. There is hope that such an approach will clarify the pathogenetic role of genes, which have so far been or will in the future be identified in candidate gene approaches or linkage studies. Such a genetic strategy may also provide the basis for diagnostic stratification of patient subgroups.

References

Aboul-Enein F, Rauschka H, Kornel B, Stadelmann C, Stefferl A, Brück W, Lucchinetti CF, Schmidbauer M, Jellinger K, Lassmann H (2003) Preferential loss of myelin associated glycoprotein reflects hypoxia-like white matter damage in stroke and inflammatory brain diseases. J Neuropath Exp Neurol 62:25–33

Archelos JJ, Hartung HP (2000) Pathogenetic role of autoantibodies in neurological disease. Trends Neurosci 23:317–327

Chang A, Tourtellotte WW, Rudick R, Trapp BD (2002) Premyelinating oligodendrocytes in chronic lesions of multiple sclerosis. New Engl J Med 346:165–173

Charcot JM (1868) Histologie de la sclerose en plaque. Gaz Hopital (Paris) 41:554–566

Charles P, Reynolds R, Seilhean D, Rougon G, Aigrot MS, Niezgoda A, Zalc B, Lubetzki C (2002) Re-expression of PSA-NCAM by demyelinated axons: an inhibitor or remyelination in multiple sclerosis? Brain 125:1972–1979

Chataway J, Sawcer S, Coraddu F, Feakes R, Broadley S, Jones HB, Clayton D, Gray J, Goodfellow PN, Compston A (1999) Evidence that allelic variants of the spinocerebellar ataxia type 2 gene influence susceptibility to multiple sclerosis. Neurogenetics 2:91–96

Compston A (2004) Genetic susceptibility and epidemiology. In: Lazzarini RA (ed) Myelin biology and disorders. Elsevier, Amsterdam, pp 701–734

Cuzner ML, Opdenakker G (1999) Plasminogen activators and matrix metalloproteinases, mediators of extracellular proteolysis in inflammatory demyelination of the central nervous system. J Neuroimmunol 94:1–14

Evangelou N, Konz D, Esiri MM, Smith S, Palace J, Matthews PM (2000a) Regional axonal loss in the corpus callosum correlates with cerebral white matter lesion volume and distribution in multiple sclerosis. Brain 123:1845–1849

Fazekas F, Strasser Fuchs S, Schmidt H, Enzinger C, Ropele S, Lechner A, Flooh E, Schmidt R, Hartung HP (2000) Apolipoprotein E genotype related differences in brain lesions of multiple sclerosis. J Neurol Neurosurg Psychiat 69:25–28

Ferguson B, Matyszak MK, Esiri MM, Perry VH (1997) Axonal damage in acute multiple sclerosis lesions. Brain 120:393–399

Franklin RJ (2002) Why does remyelination fail in multiple sclerosis? Nat Rev Neurosci 3:705–714

Giess R, Maurer M, Linker R, Gold R, Warmuth-Metz M, Toyka KV, Sendtner M, Rieckmann P (2002) Association of a null mutation in the CNTF gene with early onset of multiple sclerosis. Arch-Neurol 59:407–409

Kornek B, Lassmann H (1999) Axonal pathology in multiple sclerosis: a historical note. Brain Pathol 9:651–656

Kornek B, Storch M, Weissert R, Wallstroem E, Stefferl A, Olsson T, Linington C, Schmidbauer M, Lassmann H (2000) Multiple sclerosis and chronic autoimmune encephalomyelitis: A comparative quantitative study of axonal injury in active, inactive and remyelinated lesions. Amer J Pathol 157:267–276

Lassmann H (1998) Pathology of multiple sclerosis. In: Compston A (ed) McAlpine's Multiple Sclerosis. Churchill Liningstone, London, pp 323–358

Linington C, Bradl M, Lassmann H, Brunner C, Vass K (1988) Augmentation of demyelination in rat acute allergic encephalomyelitis by circulating antibodies against a myelin oligodendrocyte glycoprotein. Amer J Pathol 13:443–454

Linker RA, Maurer M, Gaupp S, Martini R, Holtmann B, Giess R, Rieckmann P, Lassmann H, Toyka KV, Sendtner M, Gold R (2002) CNTF is a major protective factor in demyelinating CNS disease: a neurotrophic cytokine as modulator in neuroinflammation. Nat Med 8:620–624

Lucchinetti C, Brück W, Parisi J, Scheithauer B, Rodriguez M, Lassmann H (2000) Heterogeneity of multiple sclerosis lesions: implications for the pathogenesis of demyelination. Ann Neurol 47:707–717

Lucchinetti C, Brück W, Parisi J, Scheithauer B, Rodriguez M, Lassmann H (1999) A quantitative analysis of oligodendrocytes in multiple sclerosis lesions. A study of 117 cases. Brain 122:2279–2295

Lucchinetti CF, Mandler R, McGavern D, Brück W, Gleich G, Ransohoff RM, Trebst C, Weinshenker B, Wingerchuck D, Parisi J, Lassmann H (2002) A role for humoral mechanisms in the pathogenesis of Devic's neuromyelitis optica. Brain 125:1450–1461

Mojon D, Fujihara K, Hirano M, Miller C, Lincoff N, Jacobs D, Greenberg S (1999) Leber's hereditary optic neuropathy mitochondrial DNA mutations in familial multiple sclerosis. Graefes Arch Clin Exp Ophthalmol 237:348–350

Neumann H, Medana I, Bauer J, Lassmann H (2002) Cytotoxic T lymphocytes in autoimmune and degenerative CNS diseases. Trend Neurosci 25:313–319

Niehaus A, Shi J, Grzenkowski M, Diers-Fenger M, Hartung HP, Toyka K, Bruck W, Trotter, J (2000) Patients with active relapsing-remitting multiple sclerosis synthesize antibodies recognizing oligodendrocyte progenitor cell surface protein: implications for remyelination. Ann Neurol 48:362–371

Pitt D, Werner P, Raine CS (2000) Glutamate excitotoxicity in a model of multiple sclerosis. Nat Med 6:67–70

Prineas JW, Barnard RO, Revesz T, Kwon EE, Sharer L, Cho ES(1993a) Multiple sclerosis. Pathology of recurrent lesions. Brain 116:681–693

Prineas JW, Kwon EE, Cho ES, Sharer LR, Barnett MH, Oleszak EL, Hoffman B, Morgan BP (2001) Immunopathology of secondary-progressive multiple sclerosis. Ann Neurol 50:646–657

Probert L, Eugster HP, Akassoglou K, Bauer J, Frei K, Lassmann H, Fontana A (2000) TNFR1 signalling is critical for the development of demyelination and the limitation of T-cell responses during immune-mediated CNS disease. Brain 123:2005–2019

Smith KJ, Kapoor R, Felts PA (1999) Demyelination: the role of reactive oxygen and nitrogen species. Brain Pathol 9:69–92

Smith KJ, Kapoor R, Hall SM, Davies M (2001) Electrically active axons degenerate when exposed to nitric oxide. Ann Neurol 49:470–476

Smith T, Groom A, Zhu B, Turski L (2000) Autoimmune encephalomyelitis ameliorated by AMPA antagonists. Nat Med 6:62–66

Storch MK, Piddlesden S, Haltia M, Iivanainen M, Morgan P, Lassmann H (1998) Multiple sclerosis: in situ evidence for antibody and complement mediated demyelination. Ann Neurol 43:465–471

Storch MK, Stefferl A, Brehm U, Weissert R, Wallström E, Kerschensteiner M, Olsson T, Linington C, Lassmann, H (1998) Autoimmunity to myelin oligodendrocyte glycoprotein in rats mimics the spectrum of multiple sclerosis pathology. Brain Pathol 8:681–694

Trapp BD, Peterson J, Ransohoff RM, Rudick R, Mork S, Bo L (1998) Axonal transection in the lesions of multiple sclerosis. N Engl J Med 338:278–285

Wolswijk G (2002) Oligodendrocyte precursor cells in the demyelinated multiple sclerosis spinal cord. Brain 125:338–349

3 The Neuroprotective Effect of Inflammation: Implications for the Therapy of Multiple Sclerosis

R. Hohlfeld, M. Kerschensteiner, C. Stadelmann, H. Lassmann, H. Wekerle*

3.1 Introduction

It has long been known that the immune system and the nervous system are closely linked at different levels. For example, when immune cells

---

* Additional references may be found in two review articles on which this chapter is based (Kerschensteiner et al. 2003; Hohlfeld et al. 2000).

attack the nervous system, neuroimmunological diseases arise. Multiple sclerosis (MS) and its animal models provide the paradigm for such a deleterious interaction between cells of the immune and nervous system. In experimental autoimmune encephalomyelitis (EAE) it was demonstrated that myelin-reactive T cells may have – seemingly paradoxical – neuroprotective (side-) effects (Moalem et al. 1999; Schwartz et al. 1999). So far, the precise mechanisms of these unexpected neuroprotective effects of T cells and other immune cells have remained unknown.

A different line of research has revealed that, unexpectedly, T cells and other cells of the immune system are capable of producing neurotrophic factors (Torcia et al. 1996; Kerschensteiner et al. 1999, 2003; Besser and Wank 1999). We have proposed that the two lines of investigation converge. The observed neuroprotective effects of immune cells may at least partially be mediated by their production and local secretion of neurotrophic factors (Hohlfeld et al. 2000). This concept has far-reaching consequences for the therapy of neuroimmunological diseases, especially MS.

3.2 Neurotrophic Factors

Nerve growth factor (NGF) was the first and for a long time, the only known neurotrophic factor. This small endogenous protein is best characterized by the antiapoptotic role it plays in neuronal development. Following the discovery of structurally related proteins with similar neurotrophic function in 1990, Yves Barde and Ron Lindsay coined the term "neurotrophin" for this protein family. Its members include NGF, BDNF, Neurotrophin 3 (NT 3), and Neurotrophin 4/5 (NT 4/5) (Lewin and Barde, 1996). The neurotrophins are homodimers with a conserved region containing a cysteine knot in the core of the molecule and with duplicate sites for receptor binding.

In recent years two additional families of protein growth factors that exert strong neurotrophic activity on developing neurons have been characterized. The first comprises the GDNF family ligands (GFLs). This family includes glial cell line-derived neurotrophic factor (GDNF) and the three structurally and functionally related proteins: neurturin, artemin, and persephin. All four GFLs utilize the Ret receptor tyrosine kinase as their signaling receptor. They activate Ret indirectly by binding to a family of four coreceptors referred to as GDNF α receptors (GFRα 1–4). The GFL/GFRα complexes associate tightly with Ret, forming a receptor complex in which Ret is the signal-transducing subunit.

The second family consists of the neuropoietic cytokines. Besides other more pleiotropic cytokines, it also includes the ciliary neurotrophic factor (CNTF) and leukemia inhibitory factor (LIF). While the neurotrophins and GFLs activate receptor tyrosine kinases to exert their neuroprotective activity, CNTF and LIF bind to receptors that lack intrinsic enzymatic activity. LIF binds directly to a complex of gp130 and LIF receptor β (LIFR-β). CNTF binds to the same receptor complex but indirectly via the CNTF receptor α. Activated CNTF or LIF receptors recruit cytoplasmic tyrosine kinases for their intracellular signaling.

Despite their different molecular structures and signaling mechanisms, the members of these three families of growth factors are now collectively referred to as neurotrophic factors. There are, however, other neurotrophic factors that do not belong to these families, e.g., the insulin-like growth factor-1 (IGF1).

From the viewpoint of therapy, it is important to note that the functions of neurotrophic factors are not restricted to neural development, although this aspect of the neurotrophins is best characterized. Many studies have provided solid evidence that neurotrophins also act on mature neurons, particularly on injured and degenerating nerve cells. Numerous animal models of nervous system diseases have shown that neurotrophins inhibit or delay neuronal death or degeneration. In addition, the expression levels of neurotrophins and their receptors are strongly regulated in pathological conditions. This suggests that they play a role in the response of neurons to traumatic or degenerative processes. In the healthy mature nervous system neurotrophins regulate cellular changes underlying neuronal plasticity. They trigger adaptive changes in adult neuronal morphology, modulate functional properties of synapses by both pre- and postsynaptic mechanisms, and even initiate fast synaptic responses. Furthermore, neurotrophins strongly influence the expression levels of enzymes in neurotransmitter synthesis pathways and neurotransmitter receptors. Their influence on adult central neurons, e.g., the regulation of specific neuronal properties rather than just neuronal survival, is clearly relevant for the therapy of neurodegenerative diseases.

Neurotrophins signal by binding and activating a dual receptor system consisting of receptor tyrosine kinases (the Trks) and a TNF-receptor like molecule the p75 neurotrophin receptor ($p75^{NTR}$). Signaling of the Trk receptors TrkA, TrkB, and TrkC mediates the survival-promoting effects of neurotrophins on neurons, whereas activation of $p75^{NTR}$ can trigger apoptosis. Hence, neurotrophins activate opposite cellular mechanisms depending on which type of receptor they interact with. Both the Trk and p75 receptors are expressed throughout the nervous system, and their genes are highly regulated during cell development and in adult neuronal lesions.

3.3 Evidence for Neuroprotective Functions of T Cells and Other Immune Cells

Autoimmune T cells have been shown to protect neurons from secondary degeneration after a partial crush injury of the optic nerve (Moalem et al. 1999). T cells specific for myelin basic protein (MBP), ovalbumin (OVA), or a heatshock protein (hsp) peptide were activated with their respective antigens in vitro, and then injected intraperitoneally into rats immediately after unilateral optic nerve injury. Seven days after injury, the optic nerves were analyzed immunohistochemically for the presence of T cells. Small numbers of T cells were observed in the intact (uninjured) optic nerves of rats injected with anti- MBP T cells. Greater numbers of T cells, however, were observed in the crushed optic nerves of the rats injected with T cells specific for MBP, hsp peptide, or OVA.

The degree of primary and secondary damage to the optic nerve axons and their attached retinal ganglion cells was measured by injecting a neurotracer distal to the site of the optic nerve lesion immediately after the injury, and again after two weeks (Moalem et al. 1999). The percentage of labeled retinal ganglion cells (reflecting viable axons) was significantly greater in the retinas of the rats injected with anti-MBP T cells than in the retinas of rats injected with anti-OVA or anti- hsp peptide T cells. Thus, although all three T-cell lines accumulated at the site of injury, only the MBP-specific, autoimmune T cells had a substantial effect in limiting the extent of secondary degeneration. The neuroprotective effect was confirmed by electrophysiological studies.

Taken together, these results suggest that T-cell autoimmunity can mediate significant neuroprotection after CNS injury. The authors speculate that after injury, "cryptic" epitopes might become available and might be recognized by endogenous nonencephalitogenic (benign) T cells. After local stimulation, these protective autoreactive T cells could exert their neuroprotective effect. The findings further substantiate the idea that "natural autoimmunity" can be benign and may even function as a protective mechanism (Cohen 1992).

3.4 Role of Neurotrophins in the Immune System

In view of their crucial functions in the nervous system, it was initially surprising to discover that some neurotrophins and their receptors are produced and act in the immune system (Kerschensteiner et al. 2003). NGF was the first neurotrophin shown to be expressed by immune cells (T- and B- lymphocytes, macrophages, and mast cells). B-lymphocytes also express the two NGF receptors, $p75^{NTR}$ and TrkA, and can therefore react to NGF stimulation. Indeed, important functions of B cells such as proliferation, immunoglobulin production, and cell survival are regulated by NGF. Different functions of macrophages, e.g., antigen presentation and migration, are also influenced by NGF. Moreover, a recent study showed that airway inflammation and hyperresponsiveness are disturbed in $p75^{NTR}$ gene-deficient mice, thus illustrating the importance of $p75^{NTR}$ signaling for immune regulation.

More recently, we and others have described the expression of BDNF, another member of the NGF neurotrophin family, in immune cells (Kerschensteiner et al. 2003). BDNF, which was thought to be primarily expressed in the nervous system, can be produced by essentially all major cell types of the human peripheral immune system, including $CD4^+$ and $CD8^+$ T- lymphocytes, B-lymphocytes, and monocytes in vitro. In addition, BDNF production was demonstrated in infiltrating immune cells in situ in experimental models of CNS damage and airway inflammation (Kerschensteiner et al. 2003). As there are also first reports indicating the presence of NT 3 and NT 4/5 in immune cells, one may conclude that most, if not all, members of the NGF neurotrophin family can be produced within the immune system. It is conceivable that immune cells can also be the target of auto- or paracrine neurotrophin actions, because they express some of the neurotrophin receptors. It therefore appears likely that neurotrophins can mediate bidirectional cross-talk between the nervous and immune systems (Kerschensteiner et al. 2003).

3.5 Neurotrophin Expression in MS and Other Neurological Diseases

BDNF is the first member of the neurotrophins known to be expressed in inflammatory brain lesions of MS patients (Kerschensteiner et al. 2003). In line with findings of in vitro studies, BDNF expression in MS is not restricted to neuronal cell populations. Conspicuous BDNF immunoreactivity was also demonstrated in infiltrating immune cells, especially T cells and macrophages, as well as in neurons and reactive astrocytes (Kerschensteiner et al. 2003). In particular, a larger proportion of immune cells express BDNF in actively demyelinating areas of MS lesions than in areas without ongoing myelin breakdown. Outside MS lesions, neurons are the major source of BDNF. It is well established that BDNF can be anterogradely transported and released by neurons. This process is upregulated after axonal injury and transection. Its common occurrence in MS lesions suggests that neuronal BDNF might also contribute to endogenous neurotrophic support in MS lesions.

Most of the literature agrees that neurons are the major targets for neurotrophic interactions in the CNS. In particular, the full-length isoforms of TrkB (receptor for BDNF and NT 4/5) and TrkC (receptor for NT 3) generally occur on neuronal cells. Likewise, full-length TrkB (gp145TrkB) is prominently expressed in neurons in the immediate vicinity of MS plaques (Kerschensteiner et al. 2003). Single neurons with pronounced gp145TrkB immunoreactivity are also observed close to MS lesions, suggesting that TrkB is upregulated in a proportion of damaged neurons.

In contrast, TrkA (the receptor for NGF) is expressed only in inflammatory cells in MS lesions. Moreover, recent studies on the expression pattern of the p75 neurotrophin receptor indicate that besides oligodendrocytes and oligodendrocyte precursors, macrophages and microglia are the major $p75^{NTR}$-expressing cells in MS lesions. Thus, NGF signaling via $p75^{NTR}$ or TrkA is likely to play a role in neurotrophin-mediated immunomodulation.

An important feature of the endogenous expression of neurotrophins in MS lesions is that most neurotrophins seem to be present at the actively demyelinating edge of the lesion early in its development. They are released just adjacent to axons, which may not be directly attacked by inflammatory cells, but are still at high risk of bystander damage. Thus, it is possible that a "penumbra" of subtotal damage exists around the MS plaque. A delicate balance between destructive and protective factors may prevail in this area. Neuroprotective strategies in MS could, therefore, focus on rescuing these axons, e.g., by reinforcing endogenous neurotrophic support in the periplaque area.

Fewer endogenous neurotrophins are present in older, chronic MS plaques than in the early stages of lesion development. This may be one reason for the ongoing axonal degeneration in these plaques in the chronic progressive, "neurodegenerative" stage of the disease. A reasonable strategy for slowing axonal degeneration in this late phase of MS, which is notoriously resistant to other forms of therapy, would be to provide exogenous (therapeutic) neurotrophic support.

Neurotrophic support by infiltrating immune cells most likely occurs in other conditions as well. Studies in experimental models of ischemic, traumatic, or degenerative CNS disorders describe the production of BDNF and other members of the NGF neurotrophin family by infiltrating immune cells. It is noteworthy that activated microglia can serve as an additional source of neurotrophins, both in culture and in human CNS disease such as human immunodeficiency virus type 1 encephalitis. Notably, BDNF and GDNF expression are also increased in infiltrating macrophages and local microglia in a striatal injury model. This is in striking contrast to the generally reduced expression of neurotrophins by resident CNS cells during neurodegenerative CNS diseases like Parkinson's and Alzheimer's disease. Since a reduction of autocrine trophic support is thought to be involved in the pathogenesis of neurodegeneration, one can speculate that immune cell-derived neurotrophic support may represent a compensatory mechanism.

3.6 Protective Function of Endogenous Neurotrophic Factors

Most of the established and experimental MS therapies attempt to suppress or modulate the dysregulated immune system. Practical strategies to protect or repair the jeopardized myelin-axonal unit are rare, although the various neurotrophic factors appear to be prime candidates for such strategies. Recent studies suggest that endogenous neurotrophic factors protect the myelin-axonal unit from an inflammatory assault – a finding that may open new possibilities for MS therapy. These studies employed experimental autoimmune encephalomyelitis (EAE) as an animal model of MS (Linker et al. 2002; Butzkueven et al. 2002).

Linker and colleagues induced chronic-relapsing EAE in mice that were genetically deficient for the neurotrophic factor ciliary neurotrophic factor (CNTF; Linker et al. 2002). Compared with wild-type mice, CNTF-deficient (knock-out) mice developed more severe clinical deficits and had an earlier disease onset and poorer recovery. These findings are partially reflected in MS patients who have naturally occurring null mutations of the CNTF gene, in whom disease onset also occurs earlier and predominantly motor symptoms are reported (Giess et al. 2002).

Histopathological evaluation of CNTF-deficient mice revealed that their pronounced susceptibility to EAE was probably due to the increased vulnerability of their oligodendrocytes, rather than to any immune-system abnormality (Linker et al. 2002). The rate of oligodendrocyte apoptosis was significantly higher in CNTF-deficient mice compared with wild-type mice, suggesting that these cells are more susceptible to inflammatory assault. This finding is particularly interesting because oligodendrocyte apoptosis is believed to be an important mechanism of myelin damage, at least in a histopathologically defined subgroup of MS patients. In addition, both the total numbers of oligodendrocytes and proliferating oligodendrocyte precursor cells were significantly reduced in EAE lesions from CNTF-deficient mice, suggesting that the poorer recovery from clinical deficits may be due to the impaired remyelination by oligodendrocyte precursor cells (Linker et al. 2002). The importance of developing therapeutical strategies that target the whole myelin-axonal unit is underlined by the fact that the increased vulnerability and reduced repair capacity of oligodendrocytes in CNTF-deficient mice favor the (secondary) degeneration of axons.

Butzkueven and coworkers reported that similar alterations occurred in EAE experiments with double mutant mice, which have impaired signalling through the LIFR-β/gp130 receptor complex (Butzkueven et al. 2002). In these mice, endogenous effects of the neuropoietic cytokines CNTF and leukemia inhibitory factor (LIF) are abolished. Such a loss of endogenous CNTF and LIF signalling in the EAE model enhanced oligodendrocyte loss and, concomitantly, increased disease severity.

These studies have identified the neuropoietic cytokines CNTF and LIF as important endogenous protective mediators in experimental neuroinflammation. Interestingly, protective factors including LIF and BDNF are supplied by both resident CNS cells and infiltrating immune cells. Viewed together, these results propose new options for therapeutic intervention. Indeed, the study by Butzkueven et al. shows that daily injections of exogenous LIF can prevent oligodendrocyte loss, thereby ameliorating the disease course in a mouse model of chronic EAE (Butzkueven et al. 2002). Unlike many other therapy studies, these authors demonstrated that a benefit was also achieved when LIF administration began after the onset of clinical disease.

3.7 Neuroprotective Therapeutic Approaches for MS

In the past, neuroprotective therapies were mostly explored in neurodegenerative disorders like Parkinson's and Alzheimer's disease, and in ischemic stroke. More recently, however, neuroprotection has been proclaimed an important goal for MS therapy. The basis for widening the scope of neuroprotection is evidence that neuronal and axonal injury are key features of MS lesions (Trapp et al. 1998; Ferguson et al. 1997; Kornek and Lassmann 1999). In contrast to degenerative and ischemic

CNS injury, however, neurodegeneration in MS appears to be caused by an inflammatory, presumably autoimmune, process. The challenge for neuroprotection in MS is therefore greater than in degenerative and ischemic disorders, because MS requires the combination of neuroprotective therapy and effective immunomodulation.

The precise mechanisms of axonal and neuronal damage in MS have not been identified. According to one extreme view, MS is a primarily neurodegenerative disease. But the majority of evidence clearly supports the more traditional view, that MS is an inflammatory rather than degenerative disease. There are two broad categories of immunologically mediated neuronal injury: antigen-specific and antigen-nonspecific. In the first, T cells and antibodies (which belong to the adaptive, antigen-specific immune system) attack neurons and axons directly. In the second, inflammatory mediators of the innate, antigen- nonspecific immune system damage neurons and axons indirectly.

Much recent interest has focused on antigen-specific damage, specifically $CD8^+$ cytotoxic T-lymphocyte-mediated neurotoxic injury (Neumann et al. 2002). In a landmark study using single-cell PCR analysis of microdissected T cells, clonally expanded $CD8^+$ (cytotoxic) T cells were detected in MS brain lesions at both perivascular and intraparenchymal sites (Babbe et al. 2000). A follow-up study of the same patients found brain-infiltrating $CD8^+$ T-cell clones in the CSF and/or blood persisting for up to 5 years (Skulina et al. 2004). The nature of their target antigen(s) is unknown, but the T cells could theoretically recognize a viral (foreign) antigen or an autoantigen. Another recent study found expanded $CD8^+$ T cells in the CSF of MS patients (Jacobsen et al. 2002). Together, these results suggest that cytotoxic $CD8^+$ T cells play an important pathogenic role, at least in certain types of MS. This hypothesis is further supported by the observation that some MS lesions contain $CD8^+$ T cells that directly contact demyelinated axons and polarize their cytotoxic granules to the contact zone (Neumann et al. 2002). These $CD8^+$ T cells seem to attack neurons and axons. Neurons and axons can express MHC class I antigens and are therefore susceptible to antigen-specific lysis by $CD8^+$ T cells in vitro (Neumann et al. 2002).

In addition to these T-cell mediated mechanisms, several possibilities can be envisioned for antigen-nonspecific neuronal injury (Lassmann 2003). Activated macrophages and microglia cells are often found in close contact with severed axons in MS lesions. Their toxic effector molecules, including excitatory neurotransmitters (e.g., glutamate), proteases, and reactive nitrogen (NO) species may cause axonal injury. Interestingly, axons are especially prone to injury by NO derivatives when electrically active (Smith and Lassmann 2002). Tissue damage may occur in some lesions via a hypoxia-like injury, possibly working together with NO-mediated axonal destruction (Aboul-Enein et al. 2003). These triggering factors activate several downstream events, including membrane damage, intracellular accumulation of calcium ions and activation of Ca-dependent proteases, and eventually kill the neuron.

The theoretical scenarios of antigen-specific and antigen-nonspecific neuronal injury are complex enough, but the real-life situation could be more intricate in view of growing evidence suggesting that the inflammatory process itself has a (neuro)protective component (Kerschensteiner et al. 2003). Thus, inflammation has a contrasting nature: one side is benign and protective, while the other is malicious and toxic (Kerschensteiner et al. 2003). This poses a therapeutic dilemma as any comprehensive inflammatory suppression is likely to abolish both its destructive and protective components (Hohlfeld et al. 2000).

The neuroprotective strategies developed for degenerative and ischemic disorders cannot be simply transferred to MS. There are however, options and candidate agents for neuroprotective treatment of MS. First, it is conceivable that some established immunomodulatory treatments possess a neuroprotective component. It should be noted however, that any effects of immunomodulatory treatments on MRI features like brain atrophy, black holes, N-acetyl-aspartate (NAA) levels, etc., do not prove a genuinely "neurotrophic" effect, because these MRI changes could be secondary to anti-inflammatory and immunomodulatory actions of the drug in question. In case of glatiramer acetate (GA), it has been shown that GA-specific human T cells produce BDNF in vitro, implying that they also produce neuroprotective BDNF in vivo, providing they reach MS lesions in sufficient numbers (Ziemssen et al. 2002). Secondly, some agents first tried in neurodegenerative and ischemic disorders will be or have already been tested in MS. A pilot trial of riluzole in 16 patients with primary progressive MS indicated a favorable effect on MRI signs of axonal loss after 12 months (Kalkers et al. 2002), but further trials are needed to determine clinical benefit. Finally, there is a growing list of other agents with both immunomodulatory and neuroprotective properties. Examples include the immunophilin ligands, such as cyclosporin, tacrolimus, and sirolimus (Guo et al. 2001; Avramut and Achim 2003); and tetracycline antibiotics, such as minocycline (Brundula et al. 2002; Popovic et al. 2002).

What are the prospects for neuroprotection in MS? So far, experience with neuroprotective therapies in stroke and neurodegenerative diseases has been disappointing. Numerous sodium and calcium ion channel antagonists, N-methyl- D-aspartate-receptor antagonists, gamma-aminobutyric acid-A receptor modulators, NO-pathway modulators, free-radical scavengers and antiapoptotic agents, which showed promising results in animal stroke models, were ineffective in human trials (De Keyser et al. 1999). This dampens any over-enthusiastic expectations for MS, where the situation is complicated by the need to combine neuroprotective and immunomodulatory agents while preserving the endogenous neuroprotective potential of inflammation. However, the early and relentless progression of neuronal injury in the course of MS demands our concerted efforts to develop an efficient neuroprotective strategy, no matter how difficult it may initially appear.

3.8 Possible Therapeutic Approaches for the Future

In principle, the treatment of MS has two major objectives, namely (1) suppression of the inflammatory process, and (2) restoration and protection of glial and neuronal function. The potential neuroprotective function of inflammatory cells is relevant to both these treatment goals.

For example, it must be assumed that many of the nonselective immunosuppressive treatments that have been used for the treatment of MS eliminate the neuroprotective (benign) autoimmune cells along with the autoaggressive offenders. This may be one of the reasons why treatment with conventional immunosuppressive agents often fails to have a convincing clinical benefit (Hohlfeld 1997; Noseworthy et al. 1999). Certainly, there is a convincing rationale for immunosuppressive treatment in situations when the noxious effects of inflammation prevail. However, immunosuppressive therapy is likely to fail when the beneficial effects of the inflammatory reaction outweigh its negative consequences. In MS it is unfortunately unclear whether there is a stage of the disease when the inflammatory reaction is more beneficial than harmful.

More recent treatment strategies, most of them still experimental, attempt to modulate the autoimmune reaction selectively. The proposed treatments target various immune molecules, such as cytokines, adhesion molecules, costimulatory molecules, and last but not least, the trimolecular complex of T-cell recognition (e.g., by application of altered peptide ligands or by vaccination with T cells or T- cell receptor peptides; reviewed in Hohlfeld 1997; Hemmer et al. 2002). Autoreactive, myelin-specific T cells are known to be functionally heterogenous (Meinl et al. 1997). In the future, it will be important to learn how to preserve or even enhance the proposed neuroprotective function of the "benign" autoreactive T cells during immunomodulatory intervention.

An interesting strategy for the delivery of neuroprotective factors relies on the (retroviral) transduction of one or several neurotrophic factors into antigen- specific T cell lines (Kramer et al. 1995). As the transduced T cells are specific for an autoantigen expressed in the nervous system, they home to the sites where the relevant autoantigen is expressed, recognize their antigen and are then stimulated locally to secrete neurotrophic factor(s) (Kramer et al. 1995). The results discussed here indicate that this experimental strategy has a natural counterpart. However, it appears that the neurotrophins secreted by immune cells under natural conditions are often insufficient to prevent injury. It will therefore be worthwhile to further refine the strategies to transduce neurotrophic factors into immune cells and to exploit the homing properties of the immune cells for targeting neuroprotective factors into the nervous system (Kramer et al. 1995; Flügel et al. 1999).

Acknowledgements. The Institute for Clinical Neuroimmunology is supported by the Hermann and Lilly Schilling Foundation.

References

Aboul-Enein F, Rauschka H, Kornek B, Stadelmann C, Stefferl A, Bruck W, Lucchinetti C, Schmidbauer M, Jellinger K, Lassmann H (2003) Preferential loss of myelin-associated glycoprotein reflects hypoxia-like white matter damage in stroke and inflammatory brain diseases. J Neuropathol Exp Neurol 62:25–33

Avramut M, Achim CL (2003) Immunophilins in nervous system degeneration and regeneration. Curr Top Med Chem 3:1376–1382

Babbe H, Roers A, Waisman A, Lassmann H, Goebels N, Hohlfeld R, Friese M, Schroder R, Deckert M, Schmidt S, Ravid R, Rajewsky K (2000) Clonal expansions of CD8(+) T cells dominate the T cell infiltrate in active MS lesions as shown by micromanipulation and single cell polymerase chain reaction. J Exp Med 192:393–404

Besser M, Wank R (1999) Clonally restricted production of the neurotrophins brain-derived neurotrophic factor and neurotrophin-3 mRNA by human immune cells and Th1/Th2-polarized expression of their receptors. J Immunol 162:6303–6306

Brundula V, Rewcastle NB, Metz LM, Bernard CC, Yong VW (2002) Targeting leukocyte MMPs and transmigration: minocycline as a potential therapy for multiple sclerosis. Brain 125:1297–1308

Butzkueven H, Zhang J-G, Soilu-Hänninen M, Hochrein H, Chionh F, Shipham KA, Emery B, Turnley AM, Petratos S, Ernst M, Bartlett PF, Kilpatrick TJ (2002) LIF receptor signaling limits immune-mediated demyelination by enhancing oligodendrocyte survival. Nat Med 8:613–619

Cohen IR (1992) The cognitive paradigm and the immunological homunculus. Immunol Today 13:490–494

De Keyser J, Sulter G, Luiten PG (1999) Clinical trials with neuroprotective drugs in acute ischemic stroke: are we doing the right thing? Trends Neurosci 22:535–540

Ferguson B, Matyszak MK, Esiri MM, Perry VH (1997) Axonal damage in acute multiple sclerosis lesions. Brain 120:393–399

Flügel A, Willem M, Berkowicz T, Wekerle H (1999) Gene transfer into CD4+ T lymphocytes: Green fluorescent protein engineered, encephalitogenic T cells used to illuminate immune responses in the brain. Nat Med 5:843–847

Giess R, Maurer M, Linker R, Gold R, Warmuth-Metz M, Toyka KV, Sendtner M, Rieckmann P (2002) Association of a null mutation in the CNTF gene with early onset of multiple sclerosis. Arch Neurol 59:407–409

Guo X, Dillman JF, III, Dawson VL, Dawson TM (2001) Neuroimmunophilins: novel neuroprotective and neuroregenerative targets. Ann Neurol 50:6–16

Hemmer B, Archelos JJ, Hartung H-P (2002) New concepts in the immunopathogenesis of multiple sclerosis. Nat Rev Neurosci 3:291–301

Hohlfeld R (1997) Biotechnological agents for the immunotherapy of multiple sclerosis. Principles, problems and perspectives. Brain 120:865–916

Hohlfeld R, Kerschensteiner M, Stadelmann C, Lassmann H, Wekerle H (2000) The neuroprotective effect of inflammation: Implications for the therapy of multiple sclerosis. J Neuroimmunol 107:161–166

Jacobsen M, Cepok S, Quak E, Happel M, Gaber R, Ziegler A, Schock S, Oertel WH, Sommer N, Hemmer B (2002) Oligoclonal expansion of memory $CD8^+$ T cells in cerebrospinal fluid from multiple sclerosis patients. Brain 125:538–550

Kalkers NF, Barkhof F, Bergers E, van Schijndel R, Polman CH (2002) The effect of the neuroprotective agent riluzole on MRI parameters in primary progressive multiple sclerosis: a pilot study. Mult Scler 8:532–533

Kerschensteiner M, Gallmeier E, Behrens L, Klinkert WEF, Kolbeck R, Hoppe E, Stadelmann C, Lassmann H, Wekerle H, Hohlfeld R (1999) Activated human T cells, B cells and monocytes produce brain-derived neurotrophic factor (BDNF) in vitro and in brain lesions: A neuroprotective role of inflammation? J Exp Med 189:865–870

Kerschensteiner M, Stadelmann C, Dechant G, Wekerle H, Hohlfeld R (2003) Neurotrophic cross-talk between the nervous and immune systems: Implications for neurological diseases. Ann Neurol 53:292–304

Kornek B, Lassmann H (1999) Axonal pathology in multiple sclerosis. A historical note. Brain Pathol 9:651–656

Kramer R, Zhang Y, Gehrmann J, Gold R, Thoenen H, Wekerle H (1995) Gene transfer through the blood-nerve barrier: Nerve growth factor engineered neuritogenic T lymphocytes attenuate experimental autoimmune neuritis. Nat Med 1:1162–1166

Lassmann H (2003) Axonal injury in multiple sclerosis. J Neurol Neurosurg Psych 74:695–697

Lewin GR, Barde Y-A (1996) Physiology of the neurotrophins. Annu Rev Neurosci 19:289–317

Linker RA, Mäurer M, Gaupp S, Martini R, Holtmann B, Giess R, Rieckmann P, Lassmann H, Toyka KV, Sendtner M, Gold R (2002) CNTF is a major protective factor in demyelinating CNS disease: A neurotrophic cytokine as modulator in neuroinflammation. Nat Med 8:620–624

Meinl E, Hoch RM, Dornmair K, De Waal Malefijt R, Bontrop RE, Jonker M, Lassmann H, Hohlfeld R, Wekerle H, 't Hart BA (1997) Encephalitogenic potential of myelin basic protein-specific T cells isolated from normal rhesus macaques. Am J Pathol 150:445–453

Moalem G, Leibowitz-Amit R, Yoles E, Mor F, Cohen IR, Schwartz M (1999) Autoimmune T cells protect neurons from secondary degeneration after central nervous system axotomy. Nat Med 5:49–55

Neumann H, Medana IM, Bauer J, Lassmann H (2002) Cytotoxic T lymphocytes in autoimmune and degenerative CNS diseases. Trends Neurosci 25:313–319

Noseworthy JH, Gold R, Hartung H-P (1999) Treatment of multiple sclerosis: Recent trial and future perspectives. Curr Opin Neurol 12:279–293

Popovic N, Schubart A, Goetz BD, Zhang SC, Linington C, Duncan ID (2002) Inhibition of autoimmune encephalomyelitis by a tetracycline. Ann Neurol 51:215–223

Schwartz M, Moalem G, Leibowitz-Amit R, Cohen IR (1999) Innate and adaptive immune responses can be beneficial for CNS repair. Trends Neurosci 22:295–299

Skulina C, Schmidt S, Dornmair K, Babbe H, Roers A, Rajewsky K, Wekerle H, Hohlfeld R, Goebels N (2004) Multiple sclerosis: Brain-infiltrating CD8$^+$ T cells persist as clonal expansions in the cerebrospinal fluid and blood. Proc Natl Acad Sci 101:2428–2433

Smith KJ, Lassmann H (2002) The role of nitric oxide in multiple sclerosis. Lancet Neurology 1:232–241

Torcia M, Bracci-Laudiero L, Lucibello M, Nencioni L, Labardi D, Rubatelli A, Cozzolino F, Aloe L, Garaci E (1996) Nerve growth factor is an autocrine survival factor for memory B lymphocytes. Cell 85:345–356

Trapp BD, Peterson J, Ransohoff RM, Rudick R, Mørk S, Bø L (1998) Axonal transection in the lesion of multiple sclerosis. N Engl J Med 338:278–285

Ziemssen T, Kumpfel T, Klinkert WEF, Neuhaus O, Hohlfeld R (2002) Glatiramer acetate-specific T-helper 1-and 2-type cell lines produce BDNF: implications for multiple sclerosis therapy. Brain 125:2381–2391

4 Fibroblast Growth Factors in Oligodendrocyte Physiology and Myelin Repair

L. Decker, F. Lachapelle, L. Magy, N. Picard-Riera,
B. Nait-Oumesmar, A. Baron-Van Evercooren

4.1 Introduction

The central nervous system (CNS) has an inherent ability to generate new oligodendrocytes after myelin damage. Spontaneous myelin repair occurs in acute and chronic lesions of multiple sclerosis (MS) (Ghatak et al. 1989; Prineas et al. 1993; Storch and Lassmann 1997) as well as in animal models of induced demyelination (Bunge et al. 1961; Ludwin 1987). Although reduced in thickness, the presence of newly formed myelin correlates with restored electrophysiological conduction (Blight and Young 1989; Smith 1981), and clinical remission (Rodriguez 1997). However, the potential of endogenous remyelination in MS decreases with the disease progression (Raine et al. 1988). Several reports indicate that some MS lesions contain oligodendrocyte precursors (OPCs) and premyelinating oligodendrocytes but remain chronically demyelinated (Chang et al. 2002; Wolswijk 2002). Multiple causes could account for this failure. These include exhaustion of the pool of remyelinating cells, defects in their recruitment, absence of a permissive environment to promote their differentiation in successful myelin competent cells, and axon dysfunction (reviewed in Franklin et al. 2002).

In MS, demyelination induces not only acute conduction block but also increases the sensitivity of axons to damage by inflammatory mediators. This results in progressive axonal loss and the development of chronic and irreversible neurological deficits at later stage of the disease (Ferguson et al. 1997; Trapp et al. 1998). There is therefore, an obvious need to rebuild myelin in order to protect axons and restore their proper functions. Several approaches for myelin repair are currently being investigated. These include cell therapy and activation of endogenous remyelination. Although the feasibility of cell therapy has been demonstrated in animal models, it remains difficult to transpose in a disease with such complex features as MS (Baron-Van Evercooren and Blakemore 2004). Therefore, pharmacological approaches could circumvent many of the obstacles envisioned for cell therapy. They require the identification of the appropriate cellular target, and permissive environment. This short review presents the potential cellular targets with a particular interest on adult neural stem cell, points to one aspect of environment change that is aging, and discusses evidences for a role of fibroblast growth factor 2 (FGF-2) in promoting endogenous remyelination.

4.2 Sources of Remyelinating Cells

The development of any strategy aiming at promoting endogenous repair requires the identification of appropriate cellular targets in terms of proliferation, migration, and differentiation. Within the adult CNS, several cell types are capable of generating new oligodendrocytes in response to myelin loss. These include postmitotic oligodendrocytes and OPCs, both found at the lesion site and multipotent progenitors which are localized in discrete germinative areas of the brain and the spinal cord. While the role of postmitotic oligodendrocytes in remyelination remains elusive (reviewed in Nait-Oumesmar et al. 2000), there are several arguments indicating that OPCs play a key role in CNS remyelination. Firstly, they are scattered throughout the white and grey matter of the rodent and primate CNS (Chang et al. 2002; Gogate et al. 1994; Miller et al. 1985; Nait-Oumesmar et al. 1997; Nishiyama et al. 1996; Reynolds and Hardy 1997; Scolding et al. 1998; Wolswijk 1997, 2002). Secondly, they proliferate at a low pace in normal conditions (Horner et al. 2000), but divide and increase in cell numbers in response to demyelination (Di Bello et al. 1999; Keirstead and Blakemore 1998; Reynolds et al. 2002). Finally, retroviral labeling of this population demonstrates its involvement in myelin repair (Gensert and Goldman 1997). However, OPCs migrate poorly toward lesions, thus restricting their recruitment to areas close to the lesion (Gensert and Goldman 1997; Keirstead and Blakemore 1998). Although OPCs seem to be capable of colonizing lesions after successive events of acute demyelination (Penderis et al. 2003), they become progressively depleted in chronic lesions induced by long-term feeding with cuprizone (Mason et al. 2004), thus suggesting that exhaustion of OPCs may contribute to the chronicity of the lesion.

In addition to these committed precursors, the adult CNS harbors stem cells and their progeny which are of particular interest for CNS repair since they self- renew and can give rise to multiple lineages either in vitro or when stimulated with growth factors (Craig et al. 1996; Gritti et al. 1996). These cells are located in germinative areas such as the subventricular zone (SVZ) of the lateral ventricles and the dentate gyrus in the hippocampus in the brain, and possibly, the subependyma in the spinal cord (Weiss et al. 1996). Although progenitors of the dentate gyrus are somewhat restricted in their motility (reviewed in Gage et al. 1998), those of the adult SVZ migrate actively over long distances to the olfactory bulb where they participate in the renewal of olfactory neurons (Lois and Alvarez- Buylla 1994). These characteristics led several groups to investigate the role of neural stem cells and progenitors in neural repair, and it was clearly demonstrated that SVZ cells can be activated in response to various CNS insults and replace astrocytes or neurons according to the disease type (reviewed by Picard-Riera et al. 2004). In addition, adult SVZ cells can also generate oligodendrocytes in response to de- or dys-myelination. We showed that enhanced proliferation and migration of SVZ cells occur in response to acute focal or diffuse demyelination of the mouse brain (Nait-Oumesmar et al. 1999; Picard-Riera et al. 2002). This was also demonstrated indirectly by grafting neurospheres, primed in vitro to generate OPCs, or SVZ fragments in the dysmyelinated CNS (Lachapelle et al. 2002; Zhang et al. 1999). Moreover, intrathecal delivery of neural precursors in the brain of mice affected with experimental autoimmune encephalomyelitis (EAE) leads to improved clinical profile and reduced demyelination (Pluchino et al. 2003). Although the SVZ has been, so far, the only area explored in demyelinating conditions, the presence of other germinative areas in the rodent adult hippocampus and the spinal cord suggests that multiple sites in the CNS could generate new oligodendrocytes to contribute to myelin repair.

The OPCs and neural precursor contribution to myelin repair in animal models of demyelination led to question whether they were present in the human CNS. OPCs are found in the normal adult human CNS (Scolding et al. 1998). As stated above, some of these cells are also present in MS lesions but do not seem to proliferate nor differentiate despite the presence of healthy axons (Chang et al. 2002; Wolswijk 2002). Moreover, cells with characteristics of multipotential neural precursors were identified in the human hippocampus (Pincus et al. 1998) and the SVZ (Bernier et al. 2000; Sanai et al. 2004). While their activation in response to seizures or neurodegeneration was recently reported (Curtis et al. 2003; Hoglinger et al. 2004), their behavior in demyelinating diseases remains unexplored.

4.3 FGF-2 as a Modulator of the Oligodendrocyte Lineage

As growth factors are clearly involved in the development of the CNS and are upregulated in response to CNS insults, it seems reasonable to investigate if they could be of use to promote CNS remyelination. Several growth factors were identified as modulators of the survival, proliferation, and differentiation of the oligodendrocyte lineage. Among them, basic fibroblast growth factor (FGF-2) plays a key role in oligodendrocyte development (McKinnon et al. 1991; Richardson and Jessen 2001).

Oligodendrocytes express FGF receptors at all stages of their lineage and it seems that the different functions of FGF-2 on the oligodendrocyte lineage are orchestrated by the sequential expression of FGF-R1, FGF-R2, and FGF-R3 by oligodendroglial cells throughout their differentiation (Bansal et al. 1996). In vitro, FGF-2 stimulates survival and proliferation of OPCs (Bogler et al. 1990; Gard and Pfeiffer 1993), and proliferation of O4^{+}/GalC-pre-oligodendrocytes (Gard and Pfeiffer 1990). FGF-2 upregulates PDGF-Rα in OPCs (Bogler et al. 1990; McKinnon et al. 1990) and acts in synergy with IGF I to regulate OPC cell cycle through the MAP kinase pathway (Frederick and Wood 2004). The effect of FGF-2 on OPC proliferation may be mediated by ECM molecules such as Syndecan 3, which expression is upregulated in response to FGF-2 treatment (Diemel et al. 2003; Winkler et al. 2002). Although less potent than PDGF-A, FGF-2 triggers and modulates the migration of OPCs (Milner et al. 1997; Osterhout et al. 1997). Despite these promoting effects on the early stages of the lineage, FGF-2 is generally considered as a negative regulator of oligodendrocyte terminal differentiation. Indeed, the continuous treatment of OPCs with FGF-2 by sustaining their proliferation, stops their differentiation into mature oligodendrocytes and blocks their ability to synthesize myelin proteins (McKinnon et al. 1991). However, FGF-2 treatment in vitro leads to dedifferentiation of mature oligodendrocytes into OPCs (Bansal and Pfeiffer 1997), the noncompaction of myelin-like membranes (Fressinaud et al. 1995), as well as the inhibition of myelination in aggregate cultures (Diemel et al. 2003). These in vitro data suggest that continuous treatment with FGF-2 may have contradictory effects on remyelination, preventing the process of myelination but providing additional OPCs to repopulate demyelinated lesions. However, transient use of FGF-2 enhances the reconstitution of myelin-like membranes after chemical breakdown (Fressinaud and Vallat 1994) and inhibition of early OPC differentiation by FGF-2 treatment is prevented if FGF-2 is withdrawn (Bogler et al. 1990). In fact, transient exposure to FGF-2 enhances myelination in embryonic brain cell cultures (Magy et al. 2003). These observations should be considered when exploring the benefit of strategies based on FGF-2 delivery to the demyelinated CNS.

In vivo, the role of FGF-2 signalling during myelin development and repair is not clearly established. FGF-2 is found in the developing and adult CNS (reviewed in McMorris and McKinnon 1996) and is expressed both by neurons and astrocytes (Pettmann et al. 1986). Moreover, neurons, astrocytes, OPCs, pro- oligodendrocytes and mature oligodendrocytes express FGF-Rs differentially (Redwine et al. 1997; Yazaki et al. 1994). As in vitro, FGF-R2 and FGF-R3 are differentially regulated during oligodendrocyte maturation in vivo (Bansal et al. 2003). While injections of FGF-2 into perinatal rats increase $O4^+$ pro-oligodendrocytes (Goddard et al. 1999), it prevents myelination (Goddard et al. 2001). Surprisingly, the deletion of FGF-2 does not have obvious deleterious effect on CNS myelination (Miller et al. 2000). While FGF-R1 and FGF-R2 null mutations are lethal and provide no information on the role of FGFs in vivo, FGF-R3 null mice show a delayed onset of oligodendrocyte terminal differentiation, increased astrocyte differentiation and delayed myelination (Oh et al. 2003).

The consequences of altered expression of FGF-2 or FGF-Rs on remyelination has not been fully investigated. However, FGFs may play a role in remyelination since FGF-1, FGF-2, and FGF-Rs are transiently upregulated in various models of induced demyelination (Franklin and Hinks 1999; Messersmith et al. 2000; Tourbah et al. 1992), and CNTF seems to be an upstream regulator of FGF-2 during this event (Albrecht et al. 2003). Moreover, induced demyelination in the FGF-2 –/– mice results in increased differentiation of oligodendrocytes and improved myelination of the previously demyelinated tracks (Armstrong et al. 2002). These data are consistent with the view that prolonged exposures to FGF-2 favors OPC proliferation but delays their maturation in myelin-forming cells.

4.4 FGF-2 Enhances SVZ-Derived Oligodendrogenesis

In the adult CNS, cells of the SVZ proliferate actively and express EGF and FGF receptors. Moreover, FGF-2 stimulates the survival and proliferation of embryonic, newborn, and adult neural stem cells in vitro, and potentiates the effect of EGF on these cells (Reynolds and Weiss 1992). In view of these data, FGF-2 delivery was explored to stimulate the SVZ of the adult brain and to promote repair after brain injury. Several studies highlighted the necessity of combined growth factor treatment or specific delivery procedures to achieve this goal. Long-term intrathecal administration of FGF-2, EGF, and TGFα promoted the SVZ cell proliferation, migration, and differentiation in neurons, astrocytes, and few oligodendrocytes in the striatum (Craig et al. 1996; Fallon et al. 2000; Kuhn et al. 1997), and single peripheral injections of FGF-2 reproduced similar results (Wagner et al. 1999). We investigated the effect of one single peripheral injection of FGF-2, PDGF-AB, or EGF on the adult SVZ and its capacity to generate oligodendrocytes by grafting fragments of different regions of the SVZ of adult mice into the brain of neonate shiverer (MBP $^{shi/shi}$) or wild-type mice (Lachapelle et al. 2002). Engraftment of SVZ fragments from untreated donors generated very few oligodendrocytes. However, treatment of donors with FGF-2 or PDGF-AB prior to transplantation vigorously promoted the survival, migration, and differentiation of the grafted SVZ cells into myelin-forming oligodendrocytes, the effect being optimal with FGF-2. In contrast EGF treatment led to the formation of small tumors and had little effect on myelin formation. The effect of FGF-2 resulted most likely from increased survival, proliferation, and differentiation of cells in the SVZ since in situ analysis showed that both growth factors expanded the constitutively proliferative PSA-NCAM$^+$/nestin$^-$ and PSA-NCAM$^+$/nestin$^+$ populations of the wall of the lateral ventricle and the rostral migratory stream, and favored their differentiation toward the neuronal and oligodendroglial fate. Similar experiments were reproduced grafting FGF-2- treated adult SVZ fragments in the SVZ of adult mice 2 days after lysolecithin-induced demyelination of the corpus callosum (Fig. 1). Although FGF-2 stimulated myelination in the grafted brains compared to controls, the effect was fourfold lower than in the newborn brain, stressing the role of the environment on the ability of neural stem cells to respond to growth factors.

The mechanisms by which FGF-2 stimulates adult neural stem cell-derived oligodendrogenesis are presently unknown. However, it was recently reported that during development, FGF-2 stimulates neural stem cell-derived oligodendrogenesis (Chandran et al. 2003; Kessaris et al. 2004) and reduces the number of neurons, these phenomena being partly determined by cell lineage- specific expression of FGF-Rs (Reimers et al. 2001). The effect of FGF-2 is mediated by the activation of MAP kinase signaling and is associated with the modulation of the transcription factor

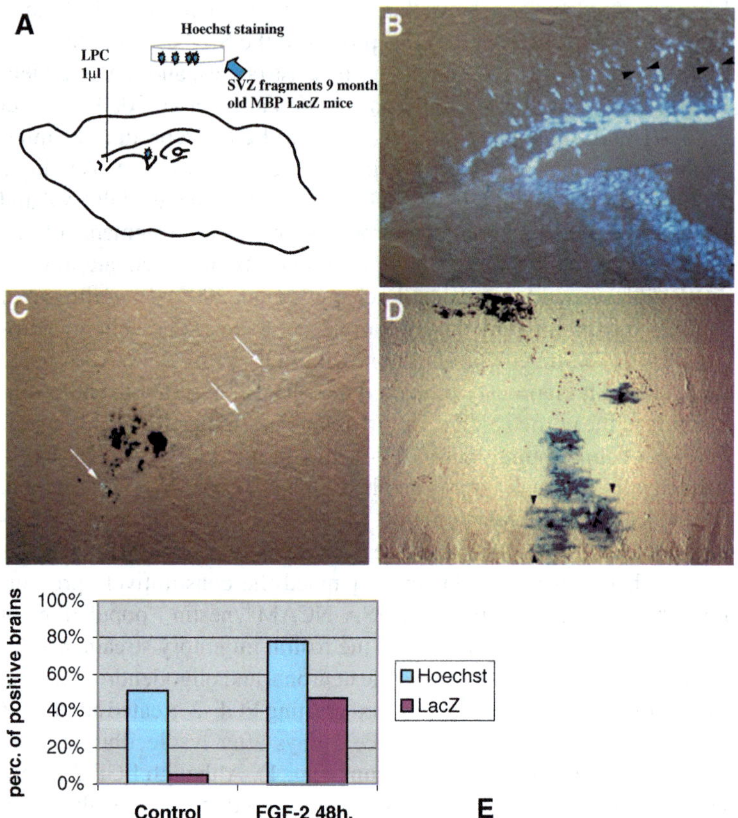

Olig2 independently of Shh signaling (Gabay et al. 2003; Kessaris et al. 2004). Future studies should indicate whether similar mechanisms are involved in neural stem cell-derived oligodendrogenesis in the adult CNS.

Fig. 1A–E. FGF-2 favors the involvement of SVZ cells in myelin repair. **A** SVZ fragments from control or FGF-2 stimulated MBP LacZ tg mice (9 months old) were Hoechst-labeled and isotopically transplanted in the SVZ of adult normal mice 48 h after LPC-induced demyelination of the corpus callosum. **B** Hoechst-labeled cells were mobilized from the SVZ into the demyelinated corpus callosum (*arrowhead*) 7 days after demyelination. **C** Some of the mobilized cells (*arrows*) reached the charcoal-labeled demyelinated area 14 days after demyelination. **D** MBP-LacZ^{+} differentiated oligodendrocytes were identified in the demyelinated area 30 days after transplantation. The arrowheads point to LacZ^{+} structures which are perpendicular to the oligodendrocyte processes and represent most likely myelin internodes. **E** FGF-2 treatment promoted a twofold increase in survival as shown by the number of Hoechst^{+} brains, and five-fold increase in differentiation of MBP-LacZ^{+} oligodendrocytes in the lesions compared to controls after transplantation of the SVZ fragments

4.5 Restricted Effect of FGF-2 in Aging CNS

Remyelination failure also occurs after experimental demyelination in aged animals (Gilson and Blakemore 1993) and seems to result from delayed remyelination (Shields et al. 1999) in correlation with altered cell proliferation, OPC recruitment, and growth factor expression (Hinks and Franklin 2000; Decker et al. 2002). Although SVZ neural stem cells and their derived progenitors maintain their capacity of proliferation throughout adult life, the proliferation of these populations decreases with age (Maslov et al. 2004; Tropepe et al. 1997). We therefore explored the effect of age on the ability of adult neural precursors to respond to lysolecithin-induced demyelination of the corpus callosum (Decker et al. 2002) and showed that, with age, the SVZ cells loose their capacity to proliferate and to be recruited by the lesion (Fig. 2).

TGFα, FGF-2, NGF, and HB-EGF infusions are capable of restoring the loss in proliferation of SVZ cells of unlesioned-aged animals to levels observed in young adults (Jin et al. 2003; Tirassa et al. 2003; Tropepe et al. 1997). We investigated the capacity of growth factor treatment to restore proliferation and recruitment of SVZ cells in aged animals after demyelination. We did find that single injections of FGF-2 or TGFα stimulated the proliferation of neural precursors located

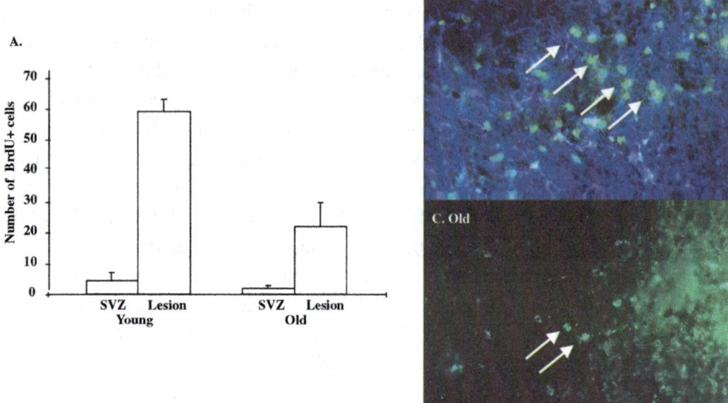

Fig. 2A–C. Age decreases the contribution of SVZ cells to colonize demyelinated area. **A** Quantification of SVZ- or RMS-traced cells 7 days after LPC-induced demyelination of the corpus callosum. BrdU injected prior the demyelination procedure was used as a cell tracer. Data revealed a decrease in cell recruitment at the lesion site with age. **B, C** Immunohistochemistry illustrating the higher number of $BrdU^+$ recruited cells (*green*) at the lesion site in young (**B**) compared with aged (**C**) mice. Note the high astrocytic reaction in the young mice (*blue*), reaction rarely detectable in old mice (not shown)

in the SVZ and RMS of young and aged mice after demyelination. However, this had little bearing on their involvement in the lesion repair since recruitment at the lesion in young mice was promoted by single injections of FGF-2 but not by TGFα, and a single injection of either factor was unable to enhance cell recruitment in aged mice. Since the migratory performances of aged neural precursors were not affected in vitro, our data suggest that aged SVZ cells may loose their intrinsic capacities to respond to demyelination-induced signals. Alternatively, due to changes in extracellular matrix molecules, reduction in glial volume and extracellular space (Sykova et al. 1998) with age, the aging brain may be nonpermissive for cell proliferation and/or recruitment in demyelinated lesions (Sim et al. 2002). Moreover, our observations stress that a more efficient mode of transient growth factor delivery should be used to induce efficient mobilization of neural precursors from the SVZ/RMS compartment to the lesioned area.

4.6 Controlled Release of FGF-2 and CNS Myelination

The above data stress the necessity of direct administration of growth factors to the CNS to optimize their efficiency but also to bypass their potential deleterious effects on other cell types of CNS and other organs. In addition, as discussed previously, continuous delivery of growth factors such as FGF-2 may exert negative effects on oligodendrocyte differentiation and thus requires a strict control in the timing of delivery in the targeted area.

Direct delivery of growth factors to the CNS resulted in improved clinical recovery and reduced demyelination in EAE models (Liu 1995; Canella 1998). Moreover, a single intrathecal injection of herpes simplex type 1 virus engineered with FGF-2 reversed significantly the neuropathological signs of EAE, increased the number of oligodendrocytes, and reduced the size of myelin lesions (Ruffini et al. 2001), suggesting that short-term FGF-2 delivery may be compatible with increased proliferation of neural progenitor and/or OPCs and their subsequent differentiation into mature oligodendrocytes. However, so far, direct delivery of growth factors remained inefficient in toxin models of CNS demyelination, suggesting that it is difficult to alter the process of remyelination by altering just one signalling molecule (O'Leary et al. 2002; Penderis et al. 2003). Implantation of fibroblasts, olfactory ensheathing cells, and Schwann cells genetically modified to release growth factors transplanted in injured spinal cord all promote axonal regeneration and OPC proliferation (Ijichi et al. 1996) and possibly myelination of regenerated axons (Blesch and Tuszynski 2003; Hendriks et al. 2004; McTigue et al. 1998; Menei et al. 1996). However, this strategy has not been used in animal models of demyelination. We developed an ex vivo strategy for gene transfer allowing the controlled release of FGF-2 by the OPC cell line CG4 (Louis et al. 1992). These cellular vectors were chosen for their ability to migrate efficiently in the adult inflammatory CNS (Tourbah et al. 1997) and capacity to differentiate in myelin-forming cells (Franklin et al. 1996; Tontsch et al. 1994). The genetically modified CG4 cells

(CG4-FGF2) were able to release significant amounts of human recombinant FGF-2 in a tetracycline- inducible fashion in an in vitro model of myelination (Magy et al. 2003). The transient release of FGF-2 by the CG4 cells enhanced the proliferation, migration, and differentiation of OPCs in an autocrine and paracrine fashion in vitro. Moreover, when the CG4-FGF 2 cells were added to mixed CNS cultures, transient exposure to FGF-2 promoted myelination. These preliminary in vitro observations suggest that short-term release of FGF-2 by CG4-FGF2 cells in a controlled manner might be of value to enhance CNS remyelination in vivo.

4.7 Conclusion

Remyelination, one of the most studied regenerative processes in the CNS, implies the proliferation, recruitment, and differentiation of oligodendrocyte lineage cells. As ensheathment of axons by newly formed myelin is capable of restoring conduction and alleviates symptoms resulting from demyelination, remyelination appears to be a promising goal in term of therapeutical approaches. FGF2 has emerged as a key factor in the early stages of oligodendrocyte development but is generally regarded as a negative factor for terminal oligodendrocyte differentiation and myelination. However, the transient expression of FGF-2 and its receptors during development and remyelination suggests that administration of this factor may be beneficial to enhance the pool of remyelinating cells, provided its release is tightly controlled to prevent its potential negative effects. In this respect, we and others have provided some evidence that transient addition of FGF-2 enhances SVZ-derived oligodendrogenesis, favors the mobilization of SVZ-derived neural precursors in vivo, and finally promotes myelination in vitro. As the sequence of growth factor expression during myelination and remyelination is highly controlled, further investigations will attempt to define the therapeutical window, allowing FGF-2 to exert its beneficial effects. Moreover, since rodent and human stem cell and progeny have different biological characteristics (Galli et al. 2000) such strategy will certainly require adjustment in the timing of pharmacological drug delivery to confirm its value in demyelinating diseases such as MS.

Acknowledgements. We are grateful to ARSEP, FRM, BERLEX, and INSERM for supporting this work. During these studies, N. Picard-Riera was supported by BERLEX and L. Decker by the Servier Institute and the FRM.

References

Albrecht PJ, Murtie JC, Ness JK, Redwine JM, Enterline JR, Armstrong RC, Levison SW (2003) Astrocytes produce CNTF during the remyelination phase of viral-induced spinal cord demyelination to stimulate FGF- 2 production. Neurobiol Dis 13:89–101

Armstrong RC, Le TQ, Frost EE, Borke RC, Vana AC (2002) Absence of fibroblast growth factor 2 promotes oligodendroglial repopulation of demyelinated white matter. J Neurosci 22:8574–8585

Bansal R, Pfeiffer SE (1997) Regulation of oligodendrocyte differentiation by fibroblast growth factors. Adv Exp Med Biol 429:69–77

Bansal R, Lakhina V, Remedios R, Tole S (2003) Expression of FGF receptors 1, 2, 3 in the embryonic and postnatal mouse brain compared with Pdgfralpha, Olig2 and Plp/dm20: implications for oligodendrocyte development. Dev Neurosci 25:83–95

Bansal R, Kumar M, Murray K, Morrison RS, Pfeiffer SE (1996) Regulation of FGF receptors in the oligodendrocyte lineage. Mol Cell Neurosci 7:263–275

Baron Van Evercooren A, Blakemore WF (2004) Remyelination through engraftment in myelin biology and disorders, vol 1, ch 6. Elsevier, San Diego, pp 143–172

Bernier PJ, Vinet J, Cossette M, Parent A (2000) Characterization of the subventricular zone of the adult human brain: evidence for the involvement of Bcl-2. Neurosci Res 37:67–78

Blesch A, Tuszynski MH (2003) Cellular GDNF delivery promotes growth of motor and dorsal column sensory axons after partial and complete spinal cord transections and induces remyelination. J Comp Neurol 467:403–417

Blight AR, Young W (1989) Central axons in injured cat spinal cord recover electrophysiological function following remyelination by Schwann cells. J Neurol Sci 91:15–34

Bogler O, Wren D, Barnett SC, Land H, Noble M (1990) Cooperation between two growth factors promotes extended self-renewal and inhibits differentiation of oligodendrocyte-type-2 astrocyte (O-2A) progenitor cells. Proc Natl Acad Sci 87:6368–6372

Bunge MB, Bunge RP, Ris H (1961) Electron microscopic study of remyelination in an experimental lesion in adult cat spinal cord. J Biophys Biochem Cytol 10:67–94

Cannella B, Hoban CJ, Gao YL, Garcia-Arenas R, Lawson D, Marchionni M, Gwynne D, Raine CS (1998) The neuregulin, glial growth factor 2, diminishes autoimmune demyelination and enhances remyelination in a chronic relapsing model for multiple sclerosis. Proc Natl Acad Sci 95(17):10100–10105

Chandran S, Kato H, Gerreli D, Compston A, Svendsen CN, Allen ND (2003) FGF-dependent generation of oligodendrocytes by a hedgehog- independent pathway. Development 130:6599–6609

Chang A, Tourtellotte WW, Rudick R, Trapp BD (2002) Premyelinating oligodendrocytes in chronic lesions of multiple sclerosis. N Engl J Med 346:165–173

Craig CG, Tropepe V, Morshead CM, Reynolds BA, S. W, van der Kooy D (1996) In vivo growth factor expansion of endogenous subependymal neural precursor cell populations in the adult mouse brain. J Neurosci 16:2649–2658

Curtis MA, Penney EB, Pearson AG, van Roon-Mom WM, Butterworth NJ, Dragunow M, Connor B, Faull RL (2003) Increased cell proliferation and neurogenesis in the adult human Huntington's disease brain. Proc Natl Acad Sci 100:9023–9027

Decker L, Picard-Riera N, Lachapelle F, Baron-Van Evercooren A (2002) Growth factor treatment promotes mobilization of young but not aged adult subventricular zone precursors in response to demyelination. J Neurosci Res 69:763–771

Di Bello IC, Dawson MR, Levine JM, Reynolds R (1999) Generation of oligodendroglial progenitors in acute inflammatory demyelinating lesions of the rat brain stem is associated with demyelination rather than inflammation. J Neurocytol 28:365–381

Diemel LT, Jackson SJ, Cuzner ML (2003) Role for TGF-beta1, FGF-2 and PDGF-AA in a myelination of CNS aggregate cultures enriched with macrophages. J Neurosci Res 74:858–867

Fallon J, Reid S, Kinyamu R, Opole I, Opole R, Baratta J, Korc M, Endo TL, Duong A, Nguyen G, Karkehabadhi M, Twardzik D, Patel S, Loughlin S (2000) In vivo induction of massive proliferation, directed migration, and differentiation of neural cells in the adult mammalian brain. Proc Natl Acad Sci 97:14686–14691

Ferguson B, Matyszak MK, Esiri MM, Perry VH (1997) Axonal damage in acute multiple sclerosis lesions. Brain 120 (Pt 3):393–399

Franklin RJ, Hinks GL (1999) Understanding CNS remyelination: clues from developmental and regeneration biology. J Neurosci Res 58:207–213

Franklin RJ, Zhao C, Sim FJ (2002) Ageing and CNS remyelination. Neuroreport 13:923–928

Franklin RJM, Bayley SA, Blakemore WF (1996) Transplanted CG4 cells (an oligodendrocyte progenitor cell line) survive, migrate, and contribute to repair of areas of demyelination in X-irradiated and damaged spinal cord but not in normal spinal cord. Exp Neurol 137:263–276

Frederick TJ, Wood TL (2004) IGF-I and FGF-2 coordinately enhance cyclin D1 and cyclin E-cdk2 association and activity to promote G1 progression in oligodendrocyte progenitor cells. Mol Cell Neurosci 25:480–492

Fressinaud C, Vallat JM (1994) Basic fibroblast growth factor improves recovery after chemically induced breakdown of myelin-like membranes in pure oligodendrocyte cultures. J Neurosci Res 38:202–213

Fressinaud C, Vallat JM, Labourdette G (1995) Basic fibroblast growth factor down-regulates myelin basic protein gene expression and alters myelin compaction of mature oligodendrocytes in vitro. J Neurosci Res 40:285–293

Gabay L, Lowell S, Rubin LL, Anderson DJ (2003) Deregulation of dorsoventral patterning by FGF confers trilineage differentiation capacity on CNS stem cells in vitro. Neuron 40:485–499

Gage FH, Kempermann G, Palmer TD, Peterson DA, Ray J (1998) Multipotent progenitor cells in the adult dentate gyrus. J Neurobiol 36:249–266

Galli R, Pagano SF, Gritti A, Vescovi AL (2000) Regulation of neuronal differentiation in human CNS stem cell progeny by leukemia inhibitory factor. Dev Neurosci 22:86–95

Gard AL, Pfeiffer SE (1990) Two proliferative stages of the oligodendrocyte lineage (A2B5+O4- and O4$^+$GalC-) under different mitogenic control. Neuron 5:615–625

Gard AL, Pfeiffer SE (1993) Glial cell mitogens bFGF and PDGF differentially regulate development of O4$^+$GalC- oligodendrocyte progenitors. Dev Biol 159:618–630

Gensert JM, Goldman JE (1997) Endogenous progenitors remyelinate demyelinated axons in the adult CNS. Neuron 19:197–203

Ghatak NR, Leshner RT, Price AC, Felton WL, 3rd (1989) Remyelination in the human central nervous system. J Neuropathol Exp Neurol 48:507–518

Gilson J, Blakemore WF (1993) Failure of remyelination in areas of demyelination produced in the spinal cord of old rats. Neuropathol Appl Neurobiol 19:173–181

Goddard DR, Berry M, Butt AM (1999) In vivo actions of fibroblast growth factor-2 and insulin-like growth factor-I on oligodendrocyte development and myelination in the central nervous system. J Neurosci Res 57:74–85

Goddard DR, Berry M, Kirvell SL, Butt AM (2001) Fibroblast growth factor-2 inhibits myelin production by oligodendrocytes in vivo. Mol Cell Neurosci 18:557–569

Gogate N, Verma L, Zhou JM, Milward E, Rusten R, O'Connor M, Kufta C, Kim J, Hudson L, Dubois-Dalcq M (1994) Plasticity in the adult human oligodendrocyte lineage. J Neurosci 14:4571–4587

Gritti A, Parati EA, Cova L, Frolichsthal P, Galli R, Wanke E, Faravelli L, Morassutti DJ, Roisen F, Nickel DD, Vescovi AL (1996) Multipotential stem cells from the adult mouse brain proliferate and self-renew in response to basic fibroblast growth factor. J Neurosci 16:1091–1100

Hendriks WT, Ruitenberg MJ, Blits B, Boer GJ, Verhaagen J (2004) Viral vector-mediated gene transfer of neurotrophins to promote regeneration of the injured spinal cord. Prog Brain Res 146:451–476

Hinks GL, Franklin RJ (2000) Delayed changes in growth factor gene expression during slow remyelination in the CNS of aged rats. Mol Cell Neurosci 16:542–556

Hoglinger GU, Rizk P, Muriel MP, Duyckaerts C, Oertel WH, Caille I, Hirsch EC (2004) Dopamine depletion impairs precursor cell proliferation in Parkinson disease. Nat Neurosci. 7(7):726–35

Horner PJ, Power AE, Kempermann G, Kuhn HG, Palmer TD, Winkler J, Thal LJ, Gage FH (2000) Proliferation and differentiation of progenitor cells throughout the intact adult rat spinal cord. J Neurosci 20:2218–2228

Ijichi A, Noel F, Sakuma S, Weil MM, Tofilon PJ (1996) Ex vivo gene delivery of platelet-derived growth factor increases 0–2A progenitors in adult rat spinal cord. Gene Ther 3:389–395

Jin K, Xie L, Childs J, Sun Y, Mao XO, Logvinova A, Greenberg DA (2003) Cerebral neurogenesis is induced by intranasal administration of growth factors. Ann Neurol 53:405–409

Keirstead HS, Blakemore WF (1998) The response of the oligodendrocyte progenitor cell population to demyelination in the adult spinal cord. Society for Neurosci 23:33–12

Kessaris N, Jamen F, Rubin LL, Richardson WD (2004) Cooperation between sonic hedgehog and fibroblast growth factor/MAPK signalling pathways in neocortical precursors. Development 131:1289–1298

Kuhn HG, Winkler J, Kempermann G, Tha LJ, Gage FH (1997) Epidermal growth factors and fibroblast growth factor-2 have different effects on neural progenitors in the adult rat brain. J Neurosci 17(15):5820–5829

Lachapelle F, Avellana-Adalid V, Nait-Oumesmar B, Baron-Van Evercooren A (2002) Fibroblast growth factor-2 (FGF-2) and platelet-derived growth factor AB (PDGF AB) promote adult SVZ-derived oligodendrogenesis in vivo. Mol Cell Neurosci 20:390–403

Liu X, Yao DL, Webster H (1995) Insulin-like growth factor I treatment reduces clinical deficits and lesion severity in acute demyelinating experimental autoimmun encephalomyelitis. Mult Scler 1:2–9

Lois C, Alvarez-Buylla A (1994) Long-distance neuronal migration in the adult mammalian brain. Science 264:1145–1148

Louis JC, Muir D, Varon S (1992) Autocrine inhibition of mitotic activity in cultured oligodendrocyte-type-2 astrocyte (O-2A) precursor cells. Glia 6:30–38

Ludwin SK (1987) Regeneration of myelin and oligodendrocytes in the central nervous system. Prog Brain Res 71:469–484

Magy L, Mertens C, Avellana-Adalid V, Keita M, Lachapelle F, Nait- Oumesmar B, Fontaine B, Baron-Van Evercooren A (2003) Inducible expression of FGF2 by a rat oligodendrocyte precursor cell line promotes CNS myelination in vitro. Exp Neurol 184:912–922

Maslov AY, Barone TA, Plunkett RJ, Pruitt SC (2004) Neural stem cell detection, characterization, and age-related changes in the subventricular zone of mice. J Neurosci 24:1726–1733

Mason JL, Toews A, Hostettler JD, Morell P, Suzuki K, Goldman JE, Matsushima GK (2004) Oligodendrocytes and progenitors become progressively depleted within chronically demyelinated lesions. Am J Pathol 164:1673–1682

McKinnon RD, Matsui T, Dubois-Dalcq M, Aaronson SA (1990) FGF modulates the PDGF-driven pathway of oligodendrocyte development. Neuron 5:603–614

McKinnon RD, Matsui T, Aranda M, Dubois-Dalcq M (1991) A role for fibroblast growth factor in oligodendrocyte development. Ann N Y Acad Sci 638:378–386

McMorris FA, McKinnon RD (1996) Regulation of oligodendrocyte development and CNS myelination by growth factors: prospects for therapy of demyelinating disease. Brain Pathol 6:313–329

McTigue DM, Horner PJ, Stokes BT, Gage FH (1998) Neurotrophin- 3 and brain-derived neurotrophic factor induce oligodendrocyte proliferation and myelination of regenerating axons in the contused adult rat spinal cord. J Neurosci 18:5354–5365

Menei P, Boisdron-Celle M, Croue A, Guy G, Benoit JP (1996) Effect of stereotactic implantation of biodegradable 5-fluorouracil-loaded microspheres in healthy and C6 glioma-bearing rats. Neurosurgery 39:117–123; discussion 123–114

Messersmith DJ, Murtie JC, Le TQ, Frost EE, Armstrong RC (2000) Fibroblast growth factor 2 (FGF2) and FGF receptor expression in an experimental demyelinating disease with extensive remyelination. J Neurosci Res 62:241–256

Miller DL, Ortega S, Bashayan O, Basch R, Basilico C (2000) Compensation by fibroblast growth factor 1 (FGF1) does not account for the mild phenotypic defects observed in FGF2 null mice. Mol Cell Biol 20:2260–2268

Miller RH, David S, Patel R, Abney ER, Raff MC (1985) A quantitative immunohistochemical study of macroglial cell development in the rat optic nerve: in vivo evidence for two distinct astrocyte lineages. Dev Biol 111:35–41

Milner R, Anderson HJ, Rippon RF, McKay JS, Franklin RJ, Marchionni MA, Reynolds R, French-Constant C (1997) Contrasting effects of mitogenic growth factors on oligodendrocyte precursor cell migration. Glia 19:85–90

Nait Oumesmar B, Vignais L, Baron Van Evercooren A (1997) Developmental expression of platelet-derived growth factor a-receptor in neurons and glial cells in the mouse CNS. J Neurosci 17:125–139

Nait-Oumesmar B, Decker L, Lachapelle F, Baron-van Evercooren A (1999) Progenitor cells of the adult subventricular zone proliferate, migrate, and differentiate into oligodendrocytes after demyelination. Eur J Neurosci 11:4357–4366

Nait-Oumesmar B, Lachapelle F, Decker L, Baron-Van Evercooren A (2000) Do central nervous system axons remyelinate? Pathol Biol (Paris) 48:70–79

Nishiyama A, Lin XH, Giese N, Heldin CH, Stallcup WB (1996) Interaction between NG2 proteoglycan and PDGF alpha-receptor on O2A progenitor cells is required for optimal response to PDGF. J Neurosci Res 43:315–330

O'Leary MT, Hinks GL, Charlton HM, Franklin RJ (2002) Increasing local levels of IGF-I mRNA expression using adenoviral vectors does not alter oligodendrocyte remyelination in the CNS of aged rats. Mol Cell Neurosci 19:32–42

Oh LY, Denninger A, Colvin JS, Vyas A, Tole S, Ornitz DM, Bansal R (2003) Fibroblast growth factor receptor 3 signaling regulates the onset of oligodendrocyte terminal differentiation. J Neurosci 23:883–894

Osterhout DJ, Ebner S, Xu J, Zazanis GA, McKinnon RD (1997) Transplanted oligodendrocyte progenitor cells expressing a dominant-negative FGF receptor transgene fail to migrate in vivo. J Neurosci 17(23):9122–9132

Penderis J, Shields SA, Franklin RJ (2003) Impaired remyelination and depletion of oligodendrocyte progenitors does not occur following repeated episodes of focal demyelination in the rat central nervous system. Brain 126:1382–1391

Pettmann B, Labourdette G, Weibel M, Sensenbrenner M (1986) The brain fibroblast growth factor (FGF) is localized in neurons. Neurosci Lett 68:175–180

Picard-Riera N, Nait-Oumesmar B, Baron-Van Evercooren A (2004) Endogenous adult neural stem cells: limits and potential to repair the injured central nervous system. J Neurosci Res 76:223–231

Picard-Riera N, Decker L, Delarasse C, Goude K, Nait-Oumesmar B, Liblau R, Pham-Dinh D, Evercooren AB (2002) Experimental autoimmune encephalomyelitis mobilizes neural progenitors from the subventricular zone to undergo oligodendrogenesis in adult mice. Proc Natl Acad Sci 99:13211–13216

Pincus DW, Keyoung HM, Harrison-Restelli C, Goodman RR, Fraser RA, Edgar M, Sakakibara S, Okano H, Nedergaard M, Goldman SA (1998) Fibroblast growth factor-2/brain-derived neurotrophic factor-associated maturation of new neurons generated from adult human subependymal cells. Ann Neurol 43:576–585

Pluchino S, Quattrini A, Brambilla E, Gritti A, Salani G, Dina G, Galli R, Del Carro U, Amadio S, Bergami A, Furlan R, Comi G, Vescovi AL, Martino G (2003) Injection of adult neurospheres induces recovery in a chronic model of multiple sclerosis. Nature 422:688–694

Prineas JW, Barnard RO, Kwon EE, Sharer LR, Cho ES (1993) Multiple sclerosis: remyelination of nascent lesions. Ann Neurol 33:137–151

Raine CS, Moore GR, Hintzen R, Traugott U (1988) Induction of oligodendrocyte proliferation and remyelination after chronic demyelination. Relevance to multiple sclerosis. Lab Invest 59:467–476

Redwine JM, Blinder KL, Armstrong RC (1997) In situ expression of fibroblast growth factor receptors by oligodendrocyte progenitors and oligodendrocytes in adult mouse central nervous system. J Neurosci Res 50:229–237

Reimers D, Lopez-Toledano MA, Mason I, Cuevas P, Redondo C, Herranz AS, Lobo MV, Bazan E (2001) Developmental expression of fibroblast growth factor (FGF) receptors in neural stem cell progeny. Modulation of neuronal and glial lineages by basic FGF treatment. Neurol Res 23:612–621

Reynolds BA, Weiss S (1992) Generation of neurons and astrocytes from isolated cells of the adult mammalian central nervous system [see comments]. Science 255:1707–1710

Reynolds R, Hardy R (1997) Oligodendroglial progenitors labeled with the O4 antibody persist in the adult rat cerebral cortex in vivo. J Neurosci 47:455–470

Reynolds R, Dawson M, Papadopoulos D, Polito A, Di Bello IC, Pham-Dinh D, Levine J (2002) The response of NG2-expressing oligodendrocyte progenitors to demyelination in MOG-EAE and MS. J Neurocytol 31:523–536

Richardson WD, Jessen KR (2001) Oligodendrocyte development. In: Glial cell development. Oxford University Press, pp 31–38

Rodriguez M (1997) Multiple sclerosis: insights into molecular pathogenesis and therapy. Mayo Clin Proc 72:663–664

Ruffini F, Furlan R, Poliani PL, Brambilla E, Marconi PC, Bergami A, Desina G, Glorioso JC, Comi G, Martino G (2001) Fibroblast growth factor-II gene therapy reverts the clinical course and the pathological signs of chronic experimental autoimmune encephalomyelitis in C57BL/6 mice. Gene Ther 8:1207–1213

Sanai N, Tramontin AD, Quinones-Hinojosa A, Barbaro NM, Gupta N, Kunwar S, Lawton MT, McDermott MW, Parsa AT, Manuel-Garcia Verdugo J, Berger MS, Alvarez-Buylla A (2004) Unique astrocyte ribbon in adult human brain contains neural stem cells but lacks chain migration. Nature 427:740–744

Scolding N, Franklin R, Stevens S, Heldin CH, Compston A, Newcombe J (1998) Oligodendrocyte progenitors are present in the normal adult human CNS and in the lesions of multiple sclerosis [see comments]. Brain 121:2221–2228

Shields SA, Gilson JM, Blakemore WF, Franklin RJ (1999) Remyelination occurs as extensively but more slowly in old rats compared to young rats following gliotoxin-induced CNS demyelination [published erratum appears in Glia 2000 Jan 1;29(1):102]. Glia 28:77–83

Sim FJ, Zhao C, Penderis J, Franklin RJ (2002) The age-related decrease in CNS remyelination efficiency is attributable to an impairment of both oligodendrocyte progenitor recruitment and differentiation. J Neurosci 22:2451–2459

Smith BH (1981) Pathophysiology of axonal transport. Neurosci Res Program Bull 20:98–106

Storch M, Lassmann H (1997) Pathology and pathogenesis of demyelinating diseases. Curr Opin Neurol 10:186–192

Sykova E, Mazel T, Simonova Z (1998) Diffusion constraints and neuron-glia interaction during aging. Exp Gerontol 33:837–851

Tirassa P, Triaca V, Amendola T, Fiore M, Aloe L (2003) EGF and NGF injected into the brain of old mice enhance BDNF and ChAT in proliferating subventricular zone. J Neurosci Res 72:557–564

Tontsch U, Archer DR, Dubois-Dalcq M, Duncan ID (1994) Transplantation of an oligodendrocyte cell line leading to extensive myelination. Proc Natl Acad Sci 91:11616–11620

Tourbah A, Baron-Van Evercooren A, Oliver L, Raulais D, Jeanny JC, Gumpel M (1992) Endogenous aFGF expression and cellular changes after a demyelinating lesion in the spinal cord of adult normal mice: immunohistochemical study. J Neurosci Res 33:47–59

Tourbah A, Linnington C, Bachelin C, Avellana-Adalid V, Wekerle H, Baron-Van Evercooren A (1997) Inflammation promotes survival and migration of the CG4 oligodendrocyte progenitors transplanted in the spinal cord of both inflammatory and demyelinated EAE rats. J Neurosci Res 50:853–861

Trapp BD, Peterson J, Ransohoff RM, Rudick R, Mork S, Bo L (1998) Axonal transection in the lesions of multiple sclerosis. N Engl J Med 338:278–285

Tropepe V, Craig CG, Morshead CM, van der Kooy D (1997) Transforming growth factor-alpha null and senescent mice show decreased neural progenitor cell proliferation in the forebrain subependyma. J Neurosci 17:7850–7859

Wagner JP, Black IB, DiCicco-Bloom E (1999) Stimulation of neonatal and adult brain neurogenesis by subcutaneous injection of basic fibroblast growth factor. J Neurosci 19:6006–6016

Weiss S, Dunne C, Hewson J, Wohl C, Wheatley M, Peterson AC, Reynolds BA (1996) Multipotent CNS stem cells are present in the adult mammalian spinal cord and ventricular neuroaxis. J Neurosci 16(23):7599–7609

Winkler S, Stahl RC, Carey DJ, Bansal R (2002) Syndecan-3 and perlecan are differentially expressed by progenitors and mature oligodendrocytes and accumulate in the extracellular matrix. J Neurosci Res 69:477–487

Wolswijk G (1997) Oligodendrocyte precursor cells in chronic multiple sclerosis lesions. Mult Scler 3:168–169

Wolswijk G (2002) Oligodendrocyte precursor cells in the demyelinated multiple sclerosis spinal cord. Brain 125:338–349

Yazaki N, Hosoi Y, Kawabata K, Miyake A, Minami M, Satoh M, Ohta M, Kawasaki T, Itoh N (1994) Differential expression patterns of mRNAs for members of the fibroblast growth factor receptor family, FGFR-1-FGFR-4, in rat brain. J Neurosci Res 37:445–452

Zhang GX, Liu TT, Ventura ES, Chen Y, Rostami A (1999) Reversal of spontaneous progressive autoimmune encephalomyelitis by myelin basic protein-induced clonal deletion. Autoimmunity 31:219–227

5 White Matter Progenitor Cells Reside in an Oligodendrogenic Niche

F.J. Sim, S.A. Goldman

5.1 The Adult Brain Harbors a Nominally Glial Progenitor

It has long been known that the adult brain contains mitotically active glia and glial progenitor cells. These parenchymal progenitor cells were first isolated from the neonatal and adult optic nerve (Raff et al. 1983; French-Constant and Raff 1986), and were first described as bipotential cells that generated oligodendrocytes and type 2 astrocytes in culture (Temple and Raff 1985). Since then, similar cells have also been isolated from all regions of the CNS (Hardy and Reynolds 1991;

Zhang et al. 1999). In vivo, the parenchymal progenitor population appears to be relatively quiescent, undergoing limited proliferation and only occasionally giving rise to mature oligodendrocytes (Gensert and Goldman 1996). These cells, commonly referred to as oligodendrocyte progenitors, are nonetheless able to respond to a variety of stimuli with increased proliferation and subsequent differentiation as oligodendrocytes (e.g., Armstrong et al. 1990; Prayoonwiwat and Rodriguez 1993; Gensert and Goldman 1997; Redwine and Armstrong 1998; Keirstead et al. 1998; Carroll et al. 1998; Sim et al. 2002). This response has in some cases led to remyelination with the formation of ultrastructurally compact myelin, restoration of saltatory conduction and locomotor function (Smith et al. 1979; Jeffery and Blakemore 1997). The importance of oligodendrocyte progenitor cells in remyelination is evidenced by the fact that ablation of these cells via high dose X-irradiation abrogates spontaneous remyelination within the irradiated fields (Blakemore and Patterson 1978; Hinks et al. 2001).

5.2 Parenchymal Progenitors Are Widespread in the Human Forebrain

In the human brain, glial progenitor pools have been found within the cortex (Arsenijevic et al. 2001) and subcortical white matter (Nunes et al. 2003). The latter cells were initially described as oligodendrocyte progenitors, on the basis of the following observations: (1) the in vitro expression of the oligodendrocyte progenitor markers PDGFαR and A2B5 (Scolding et al. 1998b, 1999); (2) the transcriptional activation of the CNP:2 promoter; (3) the generation of large numbers of $O4^+$ oligodendrocyte progeny in vitro (Roy et al. 1999); and (4) their capacity to myelinate large swathes of the shiverer mouse forebrain upon transplantation (Windrem et al. 2004), as well as to remyelinate axons following lysolecithin-induced demyelination (Windrem et al. 2002). Human parenchymal progenitor cells, whether isolated on the basis of CNP:2-promoter based sorting or A2B5 immunoreactivity, comprises 3 %–4 % of the cells of the adult white matter. They are dispersed throughout subcortical white matter, within which they appear to serve as glial progenitor cells (Scolding et al. 1999; Roy et al. 1999).

Although remyelination in experimental models is usually a robust and rapid phenomenon, remyelination is inconsistent in diseases such as multiple sclerosis (MS) (Prineas et al. 1993; Lucchinetti et al. 1996) and following spinal cord trauma (Bunge et al. 1993). The reasons for this deficiency are not yet clear (for review Franklin 2002), although the identification of immature oligodendrocytes in chronically demyelinating lesions suggests that environmental signals necessary for oligodendrocyte maturation are lacking in these foci (Wolswijk 1998; Scolding et al. 1998a; Chang et al. 2000; Maeda et al. 2001). By better understanding those environmental signals that lead to oligodendrocyte lineage commitment and maturation in the normal human brain, we hope to learn how to reactivate this response from endogenous progenitors within and surrounding demyelinating lesions.

5.3 Parenchymal Progenitors Retain a Neurogenic Capacity That Is Environmentally Suppressed

Until recently, parenchymal progenitor cells were thought capable only of producing oligodendrocytes and astrocytes. Yet multipotential neurogenic precursor cells have been isolated from the parenchyma of the adult rodent brain (Richards et al. 1992; Palmer et al. 1999; Kondo and Raff 2000). In these studies, isolated progenitors were treated with FGF2 in vitro to induce progenitor cell expansion, which was then associated with the generation of neuronal daughter cells. Kondo and Raff (2000) in particular highlighted the role of FGF2 and BMP-mediated reprogramming through an astrocyte intermediary to yield a multipotential phenotype. However, the necessity for this in vitro reprogramming stage has been questioned by recent studies. In particular, adult human white matter progenitors (WMPCs), isolated on the basis of either CNP promoter activity or A2B5 immunoselection, are intrinsically multipotent, yielding neurons as well as glia upon their initial isolation (Roy et al. 1999; Nunes et al. 2003; Belachew et al. 2003). Acutely isolated cells were able to generate functional neurons in vitro and upon immediate transplantation into the fetal rat brain (Nunes et al. 2003). Similarly, Gallo and colleagues have reported that NG2-sorted adult rodent glial progenitors are intrinsically neurogenic and multipotent (Belachew et al. 2003). These results would suggest that removal of glial progenitor cells from their local milieu may be all that is necessary to induce neurogenesis. The parenchymal progenitor population of the adult brain might thus include both multipotential and glial-restricted cellular components, dependent on their local interactions.

5.4 Parenchymal Progenitor Cells Include Multipotential Transit-Amplifying Cells

When isolated from adult human subcortical white matter, WMPCs can remain mitotically competent for several months in vitro (Nunes et al. 2003). Like other neural progenitors, WMPCs may be propagated as neurospheres, expanding as clusters of cells arising from a single parental

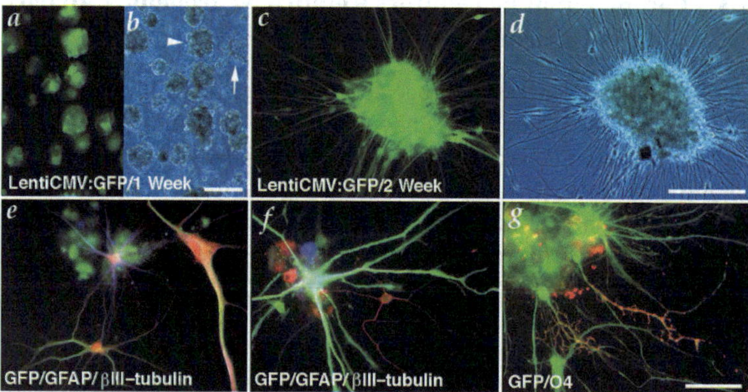

Fig. 1A–G. Single lentiviral GFP-tagged white matter progenitors generated neurons and glia. A2B5-sorted WMPCs were infected with a lentivirus encoding EGFP 5 days postsort. The primary spheres that arose in these cultures were dissociated 3 weeks later, and plated at densities of 3000 cells/per well. **A, B** The resultant secondary spheres harbored either GFP-tagged (*arrowhead*) or untagged (*arrow*) cells, and only less commonly, both. **C, D** GFP^+ secondary sphere one week after plating. **E, F** βIII-$tubulin^+$ neurons (*red*) and $GFAP^+$ astrocytes (*blue*) arising from a single GFP^+ secondary sphere. **G** GFP^+ (*green*)/$O4^+$ oligodendrocytes (*red*) arising from a secondary sphere.
Scale: **A, B** 100 μm; **C**, d 60 μm; **E–G**, 40 μm. (From Nunes et al. 2003)

cell. These spheres may be passaged, and their derived secondary and tertiary spheres retain the ability to generate neurons, oligodendrocytes, and astrocytes (Nunes et al. 2003). Lentiviral GFP tagging of WMPCs with serial passage confirmed the clonality of single multilineage spheres, in that secondary spheres either uniformly expressed GFP or not at all (Fig. 1; Nunes et al. 2003).

Although multipotential, WMPCs do not appear to be self-renewing, as might have been expected from true neural stem cells. When we assessed the extent to which WMPCs were self-renewing by repeated passage, the incidence of sphere-forming cells declined significantly with repetitive passage. In fact, WMPCs were estimated to undergo at most 18 population doublings in vitro, much less than the number expected for a tissue-derived stem cell (Nunes et al. 2003). Self-renewal capacity in the fetal human forebrain has been ascribed to sustained telomerase activity (Yashima et al. 1998; Ostenfeld et al. 2000). Consistent with their limited self-renewal capacity, WMPCs did not exhibit detectable telomerase activity (Nunes et al. 2003).

5.5 Neurons Arising from WMPCs Are Functionally Competent

Importantly, neurons generated from WMPCs were shown to develop neuronal-type calcium responses to depolarization, displaying a rapid, reversible increase in cytosolic calcium, and to generate fast sodium currents and action potentials by whole cell patch clamp analysis (Nunes et al. 2003; Fig. 2). WMPC neurogenesis was also observed in vivo following fetal xenograft transplant into the embryonic rat brain. Acutely isolated cells transplanted by intraventricular injection gave rise to all neural cell types and did so in a context-dependent manner (Nunes et al. 2003; Fig. 3). Donor-derived neurons were found in the olfactory subependyma, the rostral-migratory stream, the hippocampal alveus, and in the neostriatum, indicating an extensive capacity for neurogenesis when these cells situate in a neurogenic environment.

The neurogenic capacity of acutely isolated parenchymal progenitor cells is hardly restricted to humans. The NG2-defined glial progenitor cell can generate functional neurons in vitro when isolated from

Fig. 2A–H. Adult human WMPC-derived neurons exhibited functional maturation in vitro. **A–C** Neurons derived from white matter progenitor cells expressed a GABAergic phenotype. **A** The outgrowth of a WMPC-derived neurosphere, stained for neuronal βIII-tubulin after 35 DIV (21 days in serum-free media with FGF/NT3/PDGF, then 14 days in 2 % FBS with 20 ng/ml BDNF). **B** concurrent immunostaining for GAD67, indicating that all 9 neurons in the field are GAD67$^+$, and thus likely GABAergic neurons. **C** DAPI nuclear labeling reveals the abundance of cells in the field. **D–F** White matter progenitor cell-derived neurons developed neuronal $Ca2^+$ responses to depolarization. **D** Shows an image of WMPC-derived cells loaded with the calcium indicator dye fluo-3, 10 days after plating of first passage spheres derived from A2B5-sorted white matter (35 DIV total); many fiber-bearing cells of both neuronal and glial morphologies are apparent. **E** The same field after exposure to 100 μm glutamate, and **F** after exposure to a depolarizing stimulus of 60 mM KCl. The neurons displayed rapid, reversible, > 100 % elevations in cytosolic calcium in response to K^+, consistent with the activity of neuronal voltage-gated calcium channels. Scale = 80 m. **G–H** Whole cell patch-clamp revealed voltage-gated sodium currents and action potentials in WMPC-derived neurons. **G** A representative cell 14 days after plating of a first passage sphere derived from A2B5-sorted white matter. The cell was patch-clamped in a voltage-clamped configuration, and its responses to current injection recorded. **H** The fast negative deflections noted after current injection are typical of the voltage-gated sodium currents of mature neurons. Action potentials were noted only at $IN_a > 800$ pA. (From Nunes et al. 2003)

the postnatal rodent brain (Belachew et al. 2003). Thus, once removed from local environmental constraints, parenchymal progenitors behave as neurogenic progenitors, analogous to those found in both the subventricular zone (SVZ) and the subgranular zone of the dentate gyrus (SGZ) of the adult brain.

5.6 The Gene Expression Pattern of Adult WMPCs Suggests an Undifferentiated, Proneural Phenotype

In order to examine the molecular basis of the phenotypic plasticity of WMPCs, we chose a genomics strategy to identify those specific transcripts that distinguish the WMPC from its environment. To this end, we

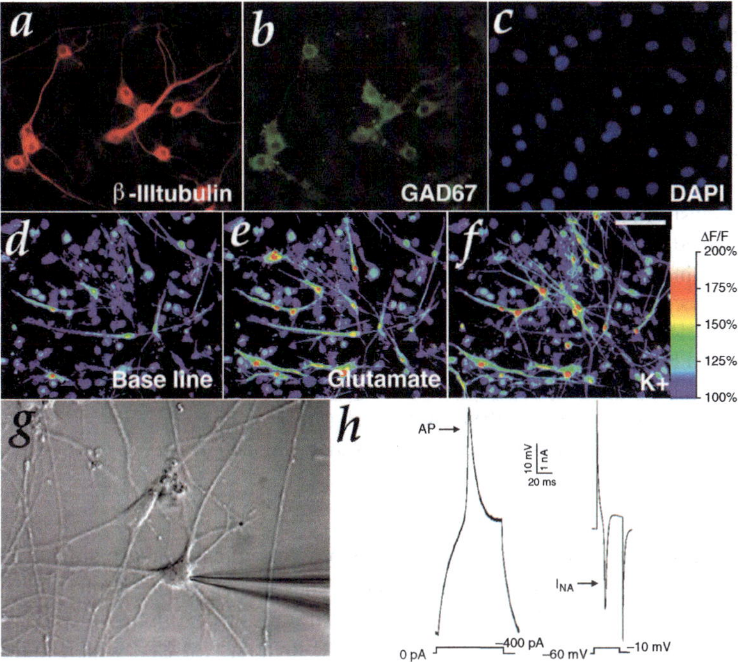

used Affymetrix microarrays to determine the transcriptional profile of A2B5-sorted human WMPCs from 4 surgically-resected adult temporal lobes. From each case, we obtained between 5×10^5 and 1×10^6 A2B5$^+$ cells that comprised roughly 3 % of all viably dissociated white matter cells. We performed microarray analysis on RNA extracted from WMPCs immediately after sorting, and RNA extracted from the matched unsorted white matter dissociate that did not undergo magnetic sorting. The expression of individual genes in each WMPC isolate was normalized against that of the unsorted white matter dissociate from which it was derived, and the mean expression ratio calculated. A set of WMPC-enriched transcripts were defined using statistical estimates of sample, per array and per gene error measurements on all genes reliably detected above background levels. This set of genes were then annotated to 210 genes using information from Affymetrix, Ensembl, and individual

Fig. 3A–H. Propagated WMPCs engrafted into fetal rats gave rise to both neurons and glia. Sections from a rat brain implanted at E17 with A2B5-sorted WMPCs. These cells were maintained in culture for 10 days prior to implantation. **A**, **B** Show nestin$^+$ (*red*) progenitors and doublecortin$^+$ (*red*) migrants, respectively, each coexpressing human nuclear antigen (hNA, *green*) in the hippocampal alveus. **C** Shows CNP$^+$ oligodendrocytes (*red*), that were found exclusively in the corpus callosum. **D** A low-power image of GFAP$^+$ (*green*, stained with anti-human GFAP) astrocytes along the ventricular wall. e βIII-tubulin$^+$ (*green*)/hNA$^+$ (*red*) neurons migrating in a chain in the hippocampal alveus. **F** βIII-tubulin$^+$ and MAP2$^+$ (inset in **F**) neurons in the striatum, adjacent to the RMS (antigens in green; hNA in red; yellow double-stained human nuclei). **G** An Hu$^+$/hNA$^+$ neuron in the septum. **H** A mature hNA$^+$ (*green*)/GAD-67$^+$ (*red*) striatal neuron. Insets in each figure show orthogonal projections of a high power confocal image of the identified cell (*arrow*).
Scale: **A–E**, 40 μm; **F–H** 20 μm. (From Nunes et al. 2003)

gene sequence data with our own custom genomics database software. Novel genes or ambiguously annotated genes were excluded from further analyses. The validity of this analysis strategy was subsequently confirmed by quantitative real time RT-PCR analysis (qPCR) which indicated that > 90 % of those 210 genes (12 of 13 genes validated) were significantly higher expression in WMPCs than their matched unsorted dissociates.

Our initial analysis examined the expression of several marker genes differentially expressed by glial progenitor cells (Fig. 4). The monoclonal antibody used to isolate adult WMPCs, clone A2B5 (Eisenbarth et al. 1979; Raff et al. 1983; Roy et al. 1999), recognizes the GQ and GT3 gangliosides and their O-acetylated derivatives (Farrer and Quarles 1999). We found that the expression of GD3 synthase, the enzyme that catalyzes the transfer of sialic acid from CMP-sialic acid to GM3, and by which GD3 and GT3 are generated, was significantly enriched in the WMPC pool. This observation was confirmed using qPCR of GD3 synthase mRNA levels following normalization to 18S ribosomal RNA. The A2B5$^+$-sorted WMPC isolates possessed high levels of PDGFR and NG2-proteoglycan expression, two canonical markers of oligodendrocyte progenitors in vivo, both of which were confirmed by

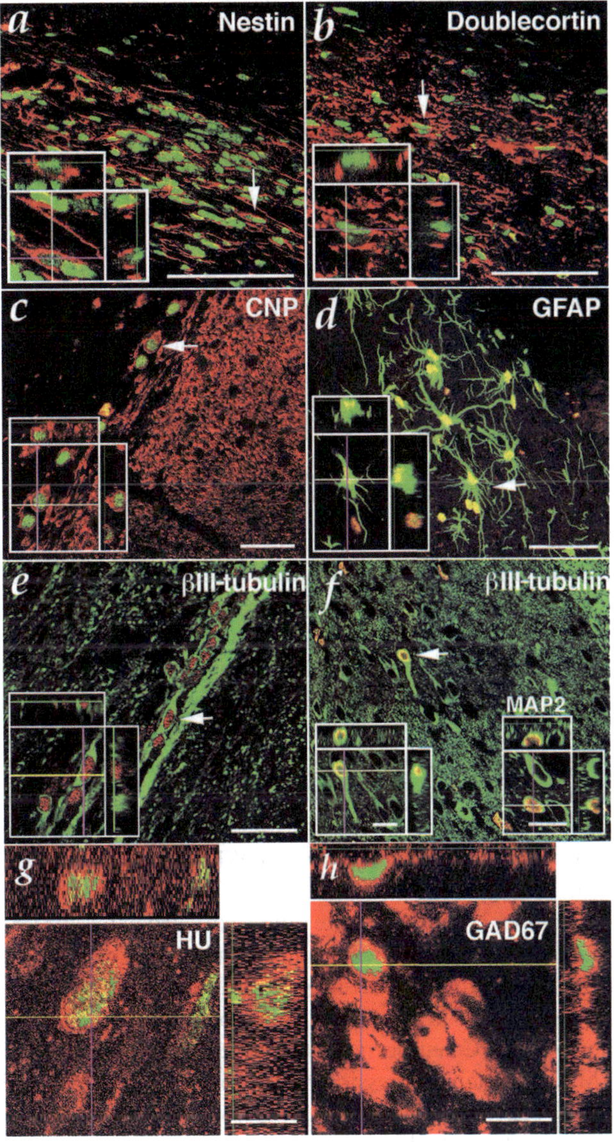
a
Nestin
b
Doublecortin
c
CNP
d
GFAP
e
βIII-tubulin
f
βIII-tubulin
MAP2
g
HU
h
GAD67

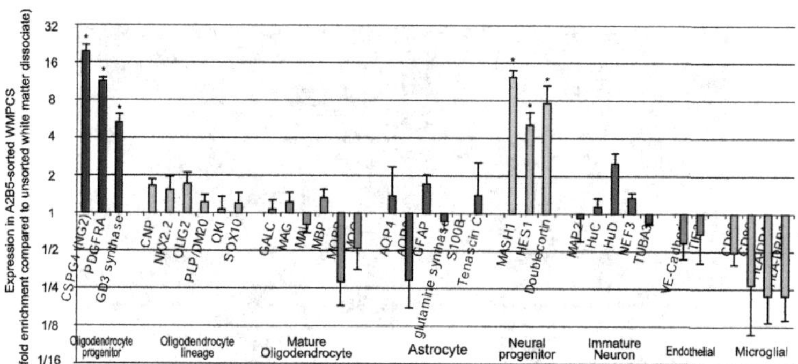

Fig. 4. WMPCs express markers of glial and neuronal progenitors. Microarray-based analysis of WMPC expressed transcripts indicated that, relative to their local tissue environment, these cells expressed the canonical oligodendrocyte progenitor markers, namely PDGFαR and NG2. Consistent with their multipotential capacity these cells also highly expressed markers of undifferentiated neural stem cells, such as HES1 and early proneural genes such as MASH1 and doublecortin

subsequent qPCR. Those transcripts restricted to more mature oligodendrocytic transcripts, including CNP and the myelin protein genes, myelin basic protein and proteolipid protein, were underexpressed by WMPCs relative to their parental white matter. Markers of other white matter phenotypes, namely astrocytes, microglia, and endothelial cells, were not highly expressed by WMPCs.

It was noteworthy that several markers characteristic of neural progenitors were differentially expressed by WMPCs. Several bHLH transcription factors typically restricted to neural progenitors and stem cells, respectively Mash1 and Hes1, were highly enriched in the WMPC pool. Both Mash1 and Hes1 are downstream components of a notch signaling pathway that has already been shown to regulate oligodendrocyte progenitor differentiation in the rat optic nerve (Wang et al. 1998) but whose regulation of WMPC neurogenic potential has yet to be determined. Recently, Mash1 protein expression has been found in a proportion of the analogous CNP-GFP/NG2-expressing rodent neurogenic progenitor (Aguirre et al. 2004). Doublecortin, which is expressed on migrating immature cells during development, was highly expressed in WMPCs. Surprisingly, GAD67 mRNA, which encodes glutamate decarboxylase and as such serves as a marker of GABA production, was greatly enriched in A2B5-sorted WMPCs. Although GABA expression has previously not been described in oligodendrocyte lineage cells, GAD expression by these cells may reflect their neurogenic potential to generate GABAergic neurons (Nunes et al. 2003; Aguirre et al. 2004). Thus, the molecular profile of WMPCs was consistent with a phenotypic plastic glial-biased neural progenitor cell.

5.7 WMPCs Are Multipotential Transit-Amplifying Cells

For a number of years, the conceptual model of tissue-specific stem cells and their progeny have been well described in skin, bone marrow, and gastrointestinal mucosae (Potten and Loeffler 1990). In this model, stem cells located in a relatively restricted niche give rise to daughter cells that, having left the stem-cell niche, become so called transit-amplifying cells. These cells have lost the ability for unlimited self-renewal, yet remain mitotically and multilineage competent, although biased and in many cases restricted to specific phenotypic lineages. These transit-amplifying cells proliferate more rapidly than stem cells so as to expand the number of terminally differentiated cells that are eventually generated.

This model has recently been applied to the adult brain and CNS, with the neuronal and glial progenitor cells considered as transit-amplifying cells derived from the stem cell located within SVZ (Doetsch et al. 2002; for review Goldman 2003; Fig. 5). The term transit-amplifying cell was first proposed for cells within the SVZ where cells were found to exhibit mitotic competence yet were biased toward a neuronal lineage (Garcia-Verdugo et al. 1998). Other neural progenitor populations located outside the SVZ may also be considered as transit-amplifying cells including the neuronally-committed progenitors of the rostral migratory stream and the neural progenitors of the subgranular zone of the dentate gyrus (Menezes et al. 1995; Doetsch et al. 2002).

On the basis of the data discussed above, WMPCs now appear to also represent a transit-amplifying parenchymal brain progenitor. They have limited self-renewal capacity, yet are capable of neurogenesis as well

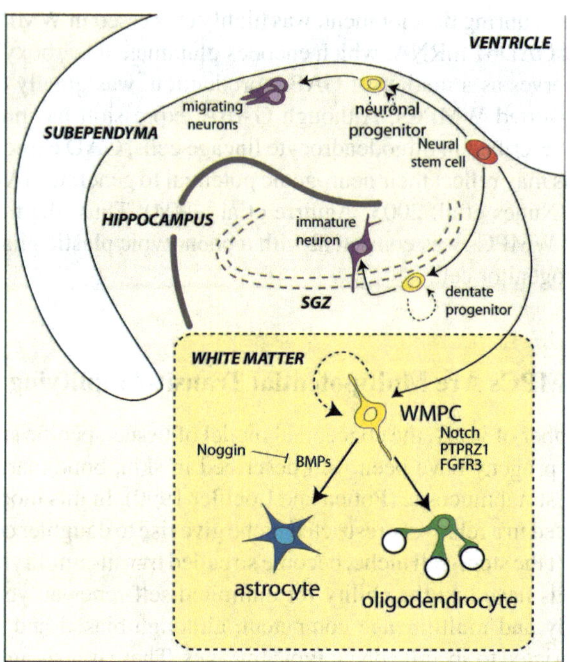

Fig. 5. Neural progenitor cell populations of the adult human brain This schematic illustrates the identified categories of progenitors in the adult human brain, and their known interrelationships. All cells are thought to derive from a common ventricular zone stem cell (*red*), which gives rise to separate neurogenic transit-amplifying cell pools (*yellow*). These include the neuronal progenitors of the ventricular subependyma, of the subgranular zone of the dentate gyrus, and of the subcortical white matter parenchyma, referred to here as white matter progenitor cells (WMPCs). These transit-amplifying cells then give rise to differentiated progeny in a context dependent manner. The WMPC appears to be regulated by a complex series of overlapping signaling pathways that determine self-renewal (Notch/PTPRZ1) and lineage commitment (BMP/Noggin). (Adapted from Goldman 2003)

as gliogenesis when removed from paracrine and autocrine signals. As such, the WMPCs resembles the multipotential neurogenic progenitors of SVZ (Morshead et al. 1994; Kirschenbaum and Goldman 1995; Weiss et al. 1996), the migrating progenitors of the rostral-migratory stream, and the subgranular zone progenitor of the dentate gyrus (SGZ) (Luskin 1993; Doetsch and Alvarez-Buylla 1996; Doetsch et al. 1997; Palmer et al. 1997; Kornack and Rakic 1999). The precise cellular relationships between the parenchymal progenitor and other neural progenitor cell populations remain to be clearly defined. Not withstanding these uncertainties, the parenchymal progenitor can now be considered a transit-amplifying cell, possessing the capabilities of limited self-renewal and multilineage competence (Nunes et al. 2003).

5.8 Does an Oligodendrogenic Niche Allow the Production of Oligodendrocytes?

Endogenous stem cells give rise to neurons when located in nominally neurogenic niches in both development and adulthood (Lim et al. 2000; Palmer et al. 2000; Louissaint, Jr. et al. 2002). The molecular environment therein has emerged as key, since even SVZ or SGZ neural progenitors when transplanted into nonneurogenic regions lose their neurogenic capacity (Lim et al. 2000; Gage 2000). On the other hand, when parenchymal progenitors are placed in the embryonic rodent brain, a permissive neurogenic niche, they are able to generate neurons in a context-dependent manner (Nunes et al. 2003). Together these observations have introduced the exciting prospect of inducing neurogenesis from parenchymal transit-amplifying cells in situ by first understanding and then manipulating the interaction of these cells with their neighbors.

One intriguing characteristic of the neurogenic niche is the proximity of neurogenic progenitors to brain blood vessels, in both the adult SVZ and SGZ (Palmer et al. 2000; Mercier et al. 2002). Angiogenesis, endothelial cell-derived BDNF, and neurogenesis appear causally related in both the adult mammalian and songbird brains (Leventhal et al. 1999; Louissaint, Jr. et al. 2002; Cao et al. 2004). Although endothelial cells can produce soluble factors that are able to stimulate self-renewal in neural stem cells (Shen et al. 2004), remarkably only those endothelial cells located within the neurogenic niche of the monkey SGZ expressed the neurotrophin BDNF (Yamashima et al. 2004). Therefore, recapitulation of this neurogenic microenvironment represents a promising strategy for inducting neurogenesis from transit-amplifying cells. Indeed a precedent has already been set in this regard, as neural progenitor cell populations of the SVZ can be induced to generate new striatal neurons following manipulation of their local environment (Kuhn et al. 1997; Benraiss et al. 2001; Pencea et al. 2001; Chmielnicki and Goldman 2002; Chmielnicki et al. 2004).

We can now propose an additional hypothesis that an analogous glial progenitor niche exists in the white matter of the adult brain that regulates both the self-renewal and competence of parenchymal progenitors to generate oligodendrocytes. Our genomics-based strategy, together with previous studies on rodent glial progenitors, has identified a number of pathways that may regulate the resting parenchymal progenitor in normal tissue, and which may become dysregulated in disease environments. For instance, the notch signaling pathway influences rodent oligodendrocyte progenitor differentiation and cell number (Wang et al. 1998; Park and Appel 2003) and a number of components of a notch pathway are expressed in human WMPC (John et al. 2002 and our unpublished observations). Also expressed in adult human WMPCs are several BMPs and their inhibitors, capable of regulating gliogenesis (Mabie et al. 1997). Indeed, inhibition of the BMP via endogenous antagonists, such as noggin, may contribute to the oligodendrogenic niche (Kondo and Raff 2004). Another candidate identified was receptor protein tyrosine phosphatase β/ζ (PTPRZ1), which is expressed on rodent oligodendrocyte progenitors in vitro, and has been shown to be necessary for recovery after demyelination (Canoll et al. 1996; Harroch et al. 2002). Finally, signaling via the FGFR3 may play an important role in generating the oligodendrogenic niche. FGFR3 expression has been shown by both rodent and human parenchymal progenitor cells (Bansal et al. 1996; Messersmith et al. 2000 and our unpublished observations). Furthermore, FGFR3 is required for timely oligodendrocyte differentiation during development (Oh et al. 2003). Intriguingly, the cognate ligands for a number of these pathways were found to be expressed, both in the white matter from which these cells were derived, and in the cells themselves, suggesting a complex pattern of both autocrine and paracrine relationships.

In summary, parenchymal progenitor cells of the adult human brain are likely maintained in an undifferentiated state by a series of parallel pathways. When removed from local and self-regulated stimuli, they become able to generate mobilized new neurons. Regenerative responses such as remyelination are thus induced when these tonically active pathways are modulated by appropriate stimuli. In disease environments such as those of chronically demyelinated foci, the oligodendrogenic niche may become dysregulated, such that the parenchyma progenitor may either remain undifferentiated or differentiate toward an undesired lineage. By perturbing the parenchymal progenitor regulatory signaling pathways in situ via pharmacologic or gene therapeutics, we may now hope to restore the oligodendrogenic niche, reactivate these pathways, and stimulate remyelination on demand.

Acknowledgements. We would like to thank our collaborators in these studies, including Drs. Neeta Roy, Martha Windrem, H. Michael Keyoung, Marta Nunes, Su Wang, and Maiken Nedergaard. Dr. Sim is supported by the Berlex Fellowship in Regenerative Medicine. This work has been supported by the NIH, the Mathers Charitable Foundation, the National Multiple Sclerosis Society, and by both Aventis Pharmaceuticals and Berlex Bioscience.

References

Aguirre AA, Chittajallu R, Belachew S, GalloV (2004) NG2- expressing cells in the subventricular zone are type C-like cells and contribute to interneuron generation in the postnatal hippocampus. J Cell Biol 165:575–589

Armstrong R, Friedrich VL, Jr., Holmes KV, Dubois-Dalcq M (1990) In vitro analysis of the oligodendrocyte lineage in mice during demyelination and remyelination. J Cell Biol 111:1183–1195

Arsenijevic Y, Villemure JG, Brunet JF, Bloch JJ, Deglon N, Kostic C, Zurn A, Aebischer P (2001) Isolation of multipotent neural precursors residing in the cortex of the adult human brain. Exp Neurol 170:48 62

Bansal R, Kumar M, Murray K, Morrison RS, Pfeiffer SE (1996) Regulation of FGF receptors in the oligodendrocyte lineage. Mol Cell Neurosci 7:263–275

Belachew S, Chittajallu R, Aguirre AA, Yuan X, Kirby M, Anderson S, Gallo V (2003) Postnatal NG2 proteoglycan-expressing progenitor cells are intrinsically multipotent and generate functional neurons. J Cell Biol 161:169–186

Benraiss A, Chmielnicki E, Lerner K, Roh D, Goldman SA (2001) Adenoviral brain-derived neurotrophic factor induces both neostriatal and olfactory neuronal recruitment from endogenous progenitor cells in the adult forebrain. J Neurosci 21:6718–6731

Blakemore WF, Patterson RC (1978) Suppression of remyelination in the CNS by X-irradiation. Acta Neuropathol 42:105–113

Bunge RP, Puckett WR, Becerra JL, Marcillo A, Quencer RM (1993) Observations on the pathology of human spinal cord injury: a review and classification of 22 new cases with details from a case of chronic compression with extensive focal demyelination. Adv Neurol 59:75–90

Canoll PD, Petanceska S, Schlessinger J, Musacchio JM (1996) Three forms of RPTP-beta are differentially expressed during gliogenesis in the developing rat brain and during glial cell differentiation in culture. J Neurosci Res 44:199–215

Cao L, Jiao X, Zuzga DS, Liu Y, Fong DM, Young D, During MJ. (2004) VEGF links hippocampal activity with neurogenesis, learning and memory. Nat Genet 36:827–835

Carroll WM, Jennings AR, Ironside LJ. (1998) Identification of the adult resting progenitor cell by autoradiographic tracking of oligodendrocyte precursors in experimental CNS demyelination. Brain 12:(Pt 2):293–302

Chang A, Nishiyama A, Peterson J, Prineas J, Trapp BD (2000) NG2- positive oligodendrocyte progenitor cells in adult human brain and multiple sclerosis lesions. J Neurosci 20:6404–6412

Chmielnicki E, Benraiss A, Economides AN, Goldman SA (2004) Adenovirally expressed noggin and brain-derived neurotrophic factor cooperate to induce new medium spiny neurons from resident progenitor cells in the adult striatal ventricular zone. J Neurosci 24:2133–2142

Chmielnicki E, Goldman SA (2002) Induced neurogenesis by endogenous progenitor cells in the adult mammalian brain. Prog Brain Res 138:451–464

Doetsch F, Alvarez-Buylla A (1996) Network of tangential pathways for neuronal migration in adult mammalian brain. Proc Natl Acad Sci 93:14895–14900

Doetsch F, Garcia-Verdugo JM, Alvarez-Buylla A (1997) Cellular composition and three-dimensional organization of the subventricular germinal zone in the adult mammalian brain. J Neurosci 17:5046–5061

Doetsch F, Petreanu L, Caille I, Garcia-Verdugo JM, Alvarez-Buylla A (2002) EGF converts transit-amplifying neurogenic precursors in the adult brain into multipotent stem cells. Neuron 36:1021–1034

Eisenbarth GS, Walsh FS, Nirenberg M (1979) Monoclonal antibody to a plasma membrane antigen of neurons. Proc Natl Acad Sci 76:4913–4917

Farrer RG, Quarles RH (1999) GT3 and its O-acetylated derivative are the principal A2B5-reactive gangliosides in cultured O2A lineage cells and are down-regulated along with O-acetyl GD3 during differentiation to oligodendrocytes. J Neurosci Res 57:371–380

FFrench-Constant C, Raff MC (1986) Proliferating bipotential glial progenitor cells in adult rat optic nerve. Nature 319:499–502

Franklin RJ. (2002) Why does remyelination fail in multiple sclerosis? Nat Rev Neurosci 3:705–714

Gage FH (2000) Mammalian neural stem cells. Science 287:1433– 1438

Garcia-Verdugo JM, Doetsch F, Wichterle H, Lim DA, Alvarez- Buylla A (1998) Architecture and cell types of the adult subventricular zone: in search of the stem cells. J Neurobiol 36:234–248

Gensert JM, Goldman JE (1996) In vivo characterization of endogenous proliferating cells in adult rat subcortical white matter. Glia 17:39– 51

Gensert JM, Goldman JE (1997) Endogenous progenitors remyelinate demyelinated axons in the adult CNS. Neuron 19:197–203

Goldman S (2003) Glia as neural progenitor cells. Trends Neurosci 26:590–596

Hardy R, Reynolds R (1991) Proliferation and differentiation potential of rat forebrain oligodendroglial progenitors both in vitro and in vivo. Development 111:1061–1080

Harroch S, Furtado GC, Brueck W, Rosenbluth J, Lafaille J, Chao M, Buxbaum JD, Schlessinger J. (2002) A critical role for the protein tyrosine phosphatase receptor type Z in functional recovery from demyelinating lesions. Nat Genet 32:411–414

Hinks GL, Chari DM, O'Leary MT, Zhao C, Keirstead HS, Blakemore WF, Franklin RJ (2001) Depletion of endogenous oligodendrocyte progenitors rather than increased availability of survival factors is a likely explanation for enhanced survival of transplanted oligodendrocyte progenitors in X-irradiated compared to normal CNS. Neuropathol Appl Neurobiol 27:59–67

Jeffery ND, Blakemore WF (1997) Locomotor deficits induced by experimental spinal cord demyelination are abolished by spontaneous remyelination. Brain 12:(Pt 1):27–37

John GR, Shankar SL, Shafit-Zagardo B, Massimi A, Lee SC, Raine CS, Brosnan CF (2002) Multiple sclerosis: Re-expression of a developmental pathway that restricts oligodendrocyte maturation. Nat Med 8:1115–1121

Keirstead HS, Levine JM, Blakemore WF (1998) Response of the oligodendrocyte progenitor cell population (defined by NG2 labelling) to demyelination of the adult spinal cord. Glia 22:161–170

Kirschenbaum B, Goldman SA (1995) Brain-derived neurotrophic factor promotes the survival of neurons arising from the adult rat forebrain subependymal zone. Proc Natl Acad Sci 92:210–214

Kondo T, Raff M (2000) Oligodendrocyte precursor cells reprogrammed to become multipotential CNS stem cells. Science 289:1754–1757

Kondo T, Raff MC (2004) A role for Noggin in the development of oligodendrocyte precursor cells. Dev Biol 267:242–251

Kornack DR, Rakic P (1999) Continuation of neurogenesis in the hippocampus of the adult macaque monkey. Proc Natl Acad Sci 96:5768–5773

Kuhn HG, Winkler J, Kempermann G, Thal LJ, Gage FH (1997) Epidermal growth factor and fibroblast growth factor-2 have different effects on neural progenitors in the adult rat brain. J Neurosci 17:5820–5829

Leventhal C, Rafii S, Rafii D, Shahar A, Goldman SA (1999) Endothelial trophic support of neuronal production and recruitment from the adult mammalian subependyma. Mol Cell Neurosci 13:450–464

Lim DA, Tramontin AD, Trevejo JM, Herrera DG, Garcia-Verdugo JM, Alvarez-Buylla A (2000) Noggin antagonizes BMP signaling to create a niche for adult neurogenesis. Neuron 28:713–726

Louissaint A, Jr., Rao S, Leventhal C, Goldman SA (2002) Coordinated interaction of neurogenesis and angiogenesis in the adult songbird brain. Neuron 34:945–960

Lucchinetti CF, Bruck W, Rodriguez M, Lassmann H (1996) Distinct patterns of multiple sclerosis pathology indicates heterogeneity in pathogenesis. Brain Pathol 6:259–274

Luskin MB (1993) Restricted proliferation and migration of postnatally generated neurons derived from the forebrain subventricular zone. Neuron 11:173–189

Mabie PC, Mehler MF, Marmur R, Papavasiliou A, Song Q, Kessler JA (1997) Bone morphogenetic proteins induce astroglial differentiation of oligodendroglial-astroglial progenitor cells. J Neurosci 17:4112–4120

Maeda Y, Solanky M, Menonna J, Chapin J, Li W, Dowling P (2001) Platelet-derived growth factor-alpha receptor-positive oligodendroglia are frequent in multiple sclerosis lesions. Ann Neurol 49:776–785

Menezes JR, Smith CM, Nelson KC, Luskin MB (1995) The division of neuronal progenitor cells during migration in the neonatal mammalian forebrain. Mol Cell Neurosci 6:496–508

Mercier F, Kitasako JT, Hatton GI (2002) Anatomy of the brain neurogenic zones revisited: fractones and the fibroblast/macrophage network. J Comp Neurol 451:170–188

Messersmith DJ, Murtie JC, Le TQ, Frost EE, Armstrong RC (2000) Fibroblast growth factor 2 (FGF2) and FGF receptor expression in an experimental demyelinating disease with extensive remyelination. J Neurosci Res 62:241–256

Morshead CM, Reynolds BA, Craig CG, McBurney MW, Staines WA, Morassutti D, Weiss S, van der KD (1994) Neural stem cells in the adult mammalian forebrain: a relatively quiescent subpopulation of subependymal cells. Neuron 13:1071–1082

Nunes MC, Roy NS, Keyoung HM, Goodman RR, McKhann G, Jiang L, Kang J, Nedergaard M, Goldman SA (2003) Identification and isolation of multipotential neural progenitor cells from the subcortical white matter of the adult human brain. Nat Med 9:439–447

Oh LY, Denninger A, Colvin JS, Vyas A, Tole S, Ornitz DM, Bansal R (2003) Fibroblast growth factor receptor 3 signaling regulates the onset of oligodendrocyte terminal differentiation. J Neurosci 23:883–894

Ostenfeld T, Caldwell MA, Prowse KR, Linskens MH, Jauniaux E, Svendsen CN (2000) Human neural precursor cells express low levels of telomerase in vitro and show diminishing cell proliferation with extensive axonal outgrowth following transplantation. Exp Neurol 164:215–226

Palmer TD, Markakis EA, Willhoite AR, Safar F, Gage FH (1999) Fibroblast growth factor-2 activates a latent neurogenic program in neural stem cells from diverse regions of the adult CNS. J Neurosci 19:8487–8497

Palmer TD, Takahashi J, Gage FH (1997) The adult rat hippocampus contains primordial neural stem cells. Mol Cell Neurosci 8:389–404

Palmer TD, Willhoite AR, Gage FH (2000) Vascular niche for adult hippocampal neurogenesis. J Comp Neurol 425:479–494

Park HC, Appel B (2003) Delta-Notch signaling regulates oligodendrocyte specification. Development 130:3747–3755

Pencea V, Bingaman KD, Wiegand SJ, Luskin MB (2001) Infusion of brain-derived neurotrophic factor into the lateral ventricle of the adult rat leads to new neurons in the parenchyma of the striatum, septum, thalamus, and hypothalamus. J Neurosci 21:6706–6717

Potten CS, Loeffler M (1990) Stem cells: attributes, cycles, spirals, pitfalls and uncertainties. Lessons for and from the crypt. Development 110:1001- -1020

Prayoonwiwat N, Rodriguez M (1993) The potential for oligodendrocyte proliferation during demyelinating disease. J Neuropathol Exp Neurol 52:55–63

Prineas JW, Barnard RO, Revesz T, Kwon EE, Sharer L, Cho E-S (1993) Multiple sclerosis: pathology of recurrent lesions. Brain 116:681–693

Raff MC, Miller RH, Noble M (1983) A glial progenitor cell that develops in vitro into an astrocyte or an oligodendrocyte depending on culture medium. Nature 303:390–396

Redwine JM, Armstrong RC (1998) In vivo proliferation of oligodendrocyte progenitors expressing PDGFαR during early remyelination. J Neurobiol 37:413–428

Richards LJ, Kilpatrick TJ, Bartlett PF (1992) De novo generation of neuronal cells from the adult mouse brain. Proc Natl Acad Sci 89:8591–8595

Roy NS, Wang S, Harrison-Restelli C, Benraiss A, Fraser RA, Gravel M, Braun PE, Goldman SA (1999) Identification, isolation, and promoter-defined separation of mitotic oligodendrocyte progenitor cells from the adult human subcortical white matter. J Neurosci 19:9986–9995

Scolding N, Franklin R, Stevens S, HELDIN CH, Compston A, Newcombe J. (1998a) Oligodendrocyte progenitors are present in the normal adult human CNS and in the lesions of multiple sclerosis. Brain 121:2221–2228

Scolding NJ, Franklin RJM., Stevens S, Heldin CH, Compston A, Newcombe J. (1998b) Oligodendrocyte progenitors are present in the normal adult human CNS and in the lesions of multiple sclerosis. Brain 121:2221–2228

Scolding NJ, Rayner PJ, Compston DA (1999) Identification of A2B5-positive putative oligodendrocyte progenitor cells and A2B5-positive astrocytes in adult human white matter. Neuroscience 89:1–4

Shen Q, Goderie SK, Jin L, Karanth N, Sun Y, Abramova N, Vincent P, Pumiglia K, Temple S (2004) Endothelial cells stimulate self-renewal and expand neurogenesis of neural stem cells. Science 304:1338–1340

Sim FJ, Zhao C, Penderis J, Franklin RJ. (2002) The age-related decrease in CNS remyelination efficiency is attributable to an impairment of both oligodendrocyte progenitor recruitment and differentiation. J Neurosci 22:2451–2459

Smith KJ, Blakemore WF, McDonald WI (1979) Central remyelination restores secure conduction. Nature 280:395–396

Temple S, Raff MC (1985) Differentiation of a bipotential glial progenitor cell in a single cell microculture. Nature 313:223–225

Wang S, Sdrulla AD, DiSibio G, Bush G, Nofziger D, Hicks C, Weinmaster G, Barres BA (1998) Notch receptor activation inhibits oligodendrocyte differentiation. Neuron 21:63–75

Weiss S, Dunne C, Hewson J, Wohl C, Wheatley M, Peterson AC, Reynolds BA (1996) Multipotent CNS stem cells are present in the adult mammalian spinal cord and ventricular neuroaxis. J Neurosci 16:7599–7609

Windrem MS, Nunes MC, Rashbaum WK, Schwartz TH, Goodman RA, McKhann G, Roy NS, Goldman SA (2004) Fetal and adult human oligodendrocyte progenitor cell isolates myelinate the congenitally dysmyelinated brain. Nat Med 10:93–97

Windrem MS, Roy NS, Wang J, Nunes M, Benraiss A, Goodman R, McKhann GM, Goldman SA (2002) Progenitor cells derived from the adult human subcortical white matter disperse and differentiate as oligodendrocytes within demyelinated lesions of the rat brain. J Neurosci Res 69:966–975

Wolswijk G (1998) Chronic stage multiple sclerosis lesions contain a relatively quiescent population of oligodendrocyte precursor cells. J Neurosci 18:601–609

Yamashima T, Tonchev AB, Vachkov IH, Popivanova BK, Seki T, Sawamoto K, Okano H (2004) Vascular adventitia generates neuronal progenitors in the monkey hippocampus after ischemia. Hippocampus 14:861

Yashima K, Maitra A, Rogers BB, Timmons CF, Rathi A, Pinar H, Wright WE, Shay JW, Gazdar AF (1998) Expression of the RNA component of telomerase during human development and differentiation. Cell Growth Differ 9:805–813

Zhang SC, Ge B, Duncan ID (1999) Adult brain retains the potential to generate oligodendroglial progenitors with extensive myelination capacity. Proc Natl Acad Sci 96:4089–4094

6 At the Interface of the Immune System and the Nervous System: How Neuroinflammation Modulates the Fate of Neural Progenitors In Vivo

F.-J. Mueller, S.R. McKercher, J. Imitola, J.F. Loring, S. Yip, S.J. Khoury, E.Y. Snyder

6.1 Introduction

The brain has recently been shown to contain stem cells that proliferate, migrate, and differentiate into glia and neurons throughout the life of the organism (Alvarez-Buylla and Lois 1995; Weiss et al. 1996; McKay 1997; Palmer et al. 1997). Despite this inherent cell replacement capacity, and the ability of progenitor cells to follow migration signals to areas of damage (Arvidsson et al. 2002; Jin et al. 2003), endogenous stem cells in the brain do not normally prove to be adequate to repair most brain injuries. Successful therapeutic interventions may entail modulation of the endogenous brain cell responses and/or the administration of stem/progenitor cells to boost the cell replacement capacity in the brain (Snyder et al. 1995; Lynch et al. 1999; Yandava et al. 1999; Aboody et al. 2000; Modo et al. 2002; Pluchino et al. 2003). For stem cell-based approaches to be successful, the poorly understood mechanisms that influence the proliferation, migration, differentiation, and integration of stem cells in the brain must be better understood.

It has been suggested that directed migration and differentiation of endogenous neural progenitor cells is one mechanism by which homeostasis is maintained day-to-day. We believe that these activities are also invoked when the organism is confronted with devastating injuries. In the most extreme cases including hypoxic-ischemic (HI) brain injury, however, it is known that in spite of mobilization of endogenous neural progenitor cells (NPCs), this response is insufficient to promote full repair and recovery. We and others have reported that transplanted neural progenitor cells are drawn to areas of damage in the brain and exert presumably beneficial effects (Flax et al. 1998; Aboody et al. 2000; Ourednik et al. 2002). In light of these observations, an important new direction in stem cell research will be focused on understanding the molecular signaling mechanisms that regulate directed migration and differentiation of stem cells.

To explain this idea, we will offer two scenarios of how pathophysiological processes could determine neural progenitor behavior. These models will provide us a tool by which the current understanding of regenerative processes within the central nervous system can be assessed and give us a basis for future research design (Figs. 1, 2).

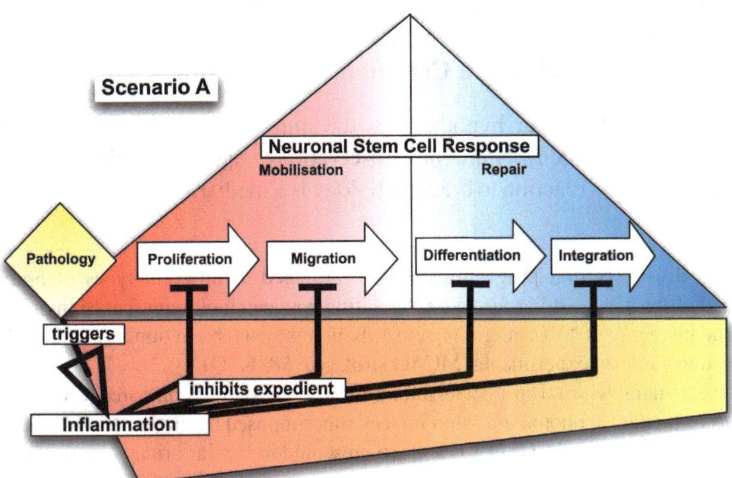

Fig. 1. Inflammation as creator of an environment detrimental for repair

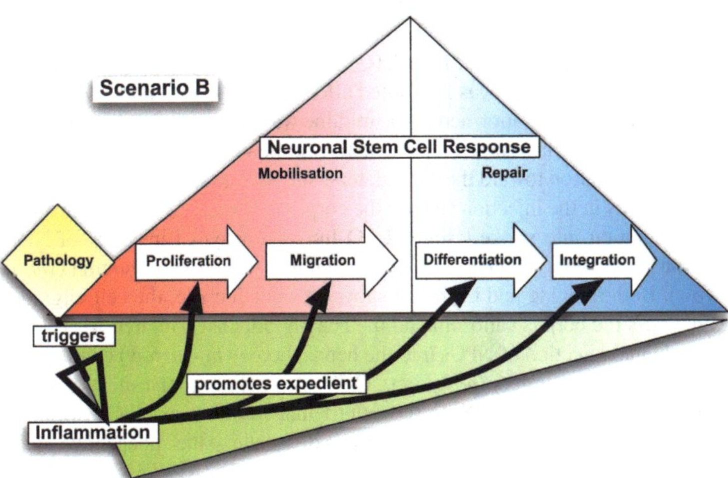

Fig. 2. Inflammation as creator of an environment permissive for repair

6.2 Is Inflammation an Important Modulator of Neural Precursor Cell Behavior? The Search For a Common Denominator

The keystone of our hypothesis is the immune response to injury in the nervous system. Numerous observations suggest that the immune system and its reaction to brain pathology is a mediator of endogenous or

Fig. 3A–Q. Neural progenitor cells are attracted to different types of brain pathologies in a similar fashion. Correlation between bioluminescence imaging and histology of NPCs migrating towards an implanted brain tumor (*p. 87*, **A–J**) and towards an experimental MCAO stroke (*p. 88*, **K–Q**).
X-gal stains with eosin counterstaining (**H–J**, **L–M**) are compared with false color luciferase photon emission images superimposed on grayscale white light anatomic images (**B–G**, **O–Q**) and are presented in similar orientation (coronal to the mouse). A X-gal expressing neural progenitor cell line (C17.2) was transfected to express the firefly luciferase gene (C17.2-LUC) without changing their stem cell properties (Tang et al. 2003).
A–J NPC migration into tumor lesions. C17.2-LUC cells were implanted into the right hemisphere of either Gli36 tumor-bearing mice (**B–D**) or control mice that did not have tumors (**E–G**). A time series of the same animal from the first group imaged on day 0 (**B**), at 1 week (**C**), and at 2 weeks (**D**). Migration toward the tumor (*dotted circle*) was first noted after 1 week (**C**; see faint bioluminescence signal) and migration across the midline was evident at 2 weeks (**D**). **E–G** Time series of another animal representative of the nontumor-bearing group, in which no migration toward the contralateral side was observed. (**H**) X-Gal- expressing cells in the injection site, (**I**) the corpus callosum, (**J**) inside the tumor. NPC migration into stroke lesions. **L**, **O** Intraparenchymal injection of NPCs into the hemisphere (*white arrow*) contralateral to the site of infarction (*yellow arrow*). Both imaging and histology show the injection site, the cell migration path across the corpus callosum and the heavily populated infarct. **M**, **P** Intraparenchymal injection of NPCs into the hemisphere (*white arrow*) contralateral to the site of sham surgery (no infarct). Cells are localized to the site of injection with no distant migration. **N**, **Q** Intraventricular injection of NPCs (*white arrow*) contralateral to the site of infarct. Cells distribute to the subependymal and periventricular areas and to the infarct. (Courtesy of Dr. R. Weissleder and his group, Center for Molecular Imaging Research, MGH, Boston. This work was first published in Tang et al. 2003 and Schellingerhout et al. 2004) (*continued on next page*)

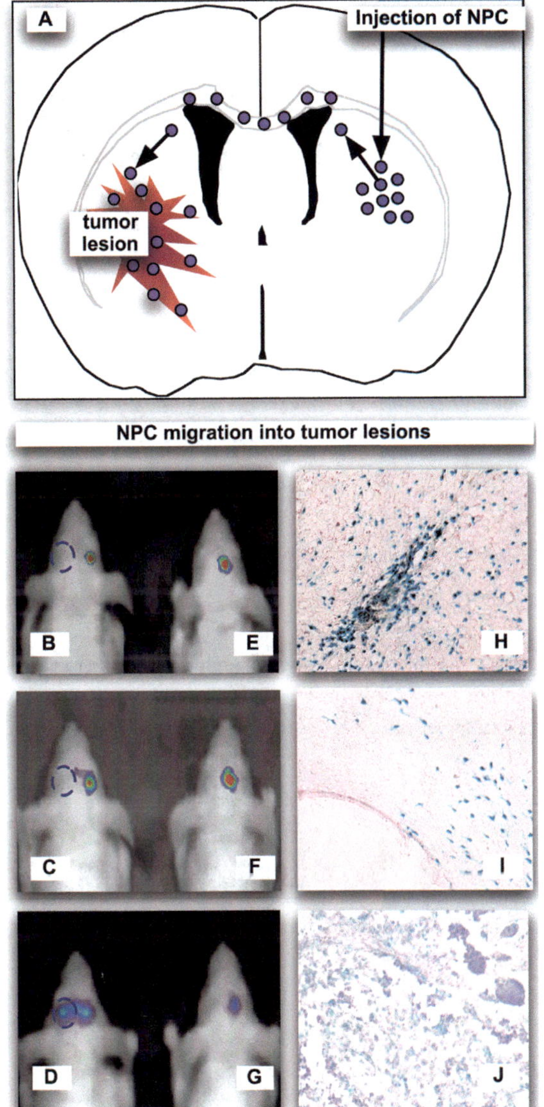
A
Injection of NPC
tumor lesion
NPC migration into tumor lesions
B
E
H
C
F
I
D
G
J

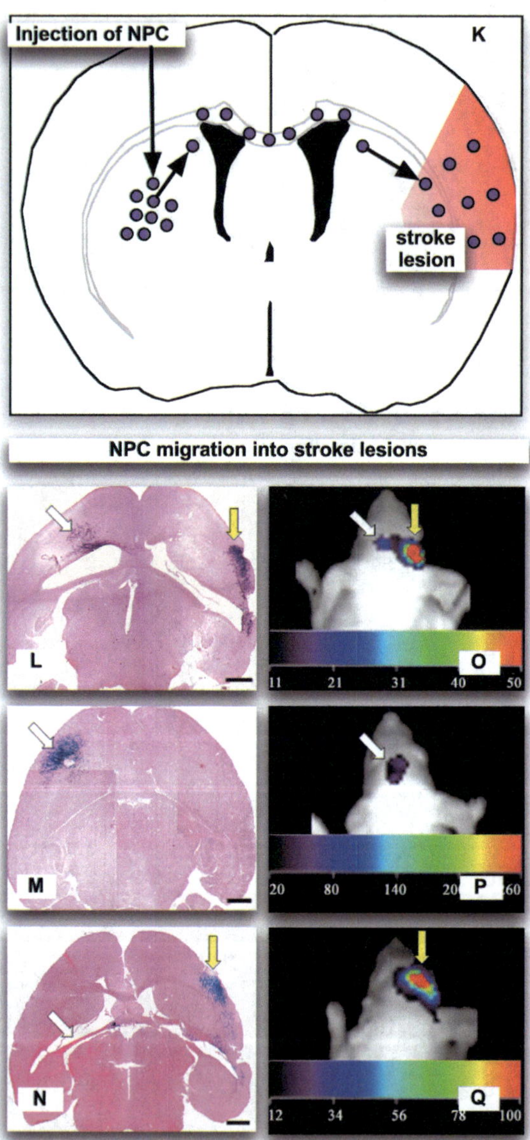

Fig. 3A–Q. (*continued*)

exogenous neural progenitor cell-based repair. It's not surprising that the primarily inflammatory pathology in animal models of multiple sclerosis (Picard-Riera et al. 2002; Pluchino et al. 2003) induces such a response. But a NPC homing response also occurs in models of acute insults such as spinal cord injury and hypoxic-ischemic brain damage (Modo et al. 2004; Zhang et al. 2004b), neoplastic changes in the brain such as gliomas (Aboody et al. 2000), genetic metabolic defects with resulting damage to the CNS (Snyder et al. 1995; Meng et al. 2003), and chronic degenerative diseases like Parkinson's disease (Ourednik et al. 2002).

The migration of progenitor cells toward areas of pathology can be exemplified by reports of exogenous neural progenitors homing into brain tumors and hypoxic-ischemic brain injury (see Fig. 3). It is becoming apparent that the one element common to all pathological conditions of the CNS that attract neural progenitor cells is that they have a prominent inflammatory component. Brain tumors such as gliomas contain up to one third inflammatory cells, (Graeber et al. 2002; Strik et al. 2004) although the immune reaction is modulated by the tumor cells (immune escape phenomenon; Hill et al. 1999). The inflammatory component in stroke is very prominent and encompasses an extreme activation of myelomonocytic cells such as microglia and invading macrophages (Han and Yenari 2003; Tan et al. 2003). The correspondence of progenitor cell migration and inflammation suggests that there may be a link between inflammation and neural progenitor cell response to pathology.

6.2.1 What Is the Nature of Inflammation in the Brain?

Inflammation and inflammatory changes in the diseased brain are not confined to cells of hematopoietic origin. Cells of mesenchymal or neuroectodermal origin also exert their influence on neural progenitors; powerful secreted factors from microglia cells activate endothelial cells and astroglia, which in turn may influence neural progenitors (Aarum et al. 2003; Liberto et al. 2004). Therefore, a useful term for the inflammatory response in the brain might be "inflammatory signature" to describe the activated state of resident brain cells in reaction to pathology. The involvement of multiple cell types implies that there might be many distinct inflammatory signatures with distinct functional features (Baker and Manuelidis 2003).

As an example, the cytokine IL-6 appears to have multiple effects that depend on other conditions in astrocytes. In transgenic mice with IL-6 overexpression in the brain, it was found that IL-6 enhances neuroprotection and neuroregeneration in models of brain insults (Loddick et al. 1998; Penkowa et al. 1999; Penkowa et al. 2003a,b et al.). However, the very same mouse line shows features of neurodegenerative disease as a result of the chronic exposure to the cytokine, consisting of neurodegeneration, blood–brain barrier breakdown, reactive gliosis, and impaired hippocampal neurogenesis (Campbell et al. 1993; Chiang et al. 1994; Brett et al. 1995; Vallieres et al. 2002). These observations point toward the possibility that the same inflammatory signature can be beneficial in an acute situation but detrimental in a chronic condition. In addition, the degree and the specific profile of various factors present in the inflamed tissue might modulate brain degeneration or regeneration.

In human neurodegenerative disorders, there appear to be different degrees of microglial activation; in some situations cells seem to spin out of control within the CNS and develop a hyperactivated state (Hailer et al. 2001; Eskes et al. 2003; Shafer et al. 2003). The reason for this distinct, eventually self-perpetuating neurodegenerative process is attributed to a grave disruption of the communication between microglia and the neuroectodermal milieu (Aldskogius and Kozlova 1998; Wada et al. 2000; Szpak et al. 2001; Polazzi and Contestabile 2002; Popovich et al. 2002; Brionne et al. 2003; Polazzi and Contestabile 2003).

6.2.2 What Does Inflammation Mean to Neural Progenitor Cells?

How do neural progenitor cells respond to inflammation? We and others have shown that neural progenitor cells express not only receptors for signals emanating from inflammatory signatures in the brain but also have the ability to reciprocally interact with and instruct cells of hematopoietic origin (Lazarini et al. 2003; Tran and Miller 2003; Imitola et al. 2004; Ji et al. 2004; Peng et al. 2004).

The two scenarios we envision (Figs. 1, 2) illustrate the interaction between pathological changes, the immune response, and neural progenitor cells. In one case, the inflammation is detrimental to stem cells; in the other, it is beneficial. These two scenarios are not mutually exclusive, but help us formulate hypotheses about what would be appropriate interventions for neurodegenerative diseases.

The neural progenitor cell response to brain damage can be divided into two main phases that have their conceptual origin in developmental neurobiology: mobilization and repair. The initial step consists of increasing the pool of neural progenitor cells and the subsequent migration of neuroblasts mainly from the subventricular zone (SVZ) to the site of injury (Arvidsson et al. 2002; Zhang et al. 2004b). At the site of injury the neuroblasts are thought to differentiate and integrate into the environment, either as oligodendrocytes, astrocytes, or neurons (Arvidsson et al. 2002). The processes of mobilization and repair could be influenced negatively or positively at every step by intervention of the immune system (Figs. 1, 2). For example, proliferation of neural progenitor cells might be increased in response to injury, but rendered insufficient because the cells are damaged by inflammatory cells. The activated immune system could intercede at any point in the process, blocking migration of neural progenitor cells, causing differentiation at the wrong site, suppressing specific lineage fates or interrupting the integration process (Fig. 1).

Conversely, inflammation could be beneficial for some phases of migration and repair. Microglia could render normally nonneurogenic brain areas into a permissive state for neurogenesis and regeneration (see Fig. 2). The progenitor proliferation might be induced by reactive astrocytes (Song et al. 2002; Liberto et al. 2004) and the migration directed by signal molecules produced by microglia or cells stimulated by microglia (Aarum et al. 2003). The differentiation could be directed by monocytic cells either by direct interaction or by creation of a neurogenic niche by inflammatory cytokines (Aarum et al. 2003; Ben-Hur et al. 2003).

These two scenarios evoke a key question: Does therapeutically modulating brain inflammation result in more or less repair?

6.3 Experimental Evidence for Functional Effects of Inflammation on Neural Precursors

6.3.1 Defining the Interface: What Signals Can Neural Precursors Understand?

Neural progenitor cells express a large variety of receptors for inflammation-related signals. Table 1 lists a number of receptors that have been shown to be expressed by neural progenitor cells or ligands to which neural progenitor cells respond. The functional importance of chemokine receptors has best been shown in vivo for CXCR4, the receptor for stromal-derived factor 1 (SDF1). A CXCR4-null mouse was reported to have severe defects in the development of the cerebellum (Ma et al. 1998; Zou et al. 1998), the dentate gyrus (Bagri et al. 2002; Lu et al. 2002), and the neocortex (Stumm et al. 2003). This is not true of all chemokine receptors; many chemokine receptor knockout mice do not display an overt abnormal brain development phenotype (Tran and Miller 2003). The difference in functional significance among receptors might be related to the degree of promiscuity these receptors show for multiple ligands (Kielian 2004). CXCR4 is unusual in that SDF1 is its only known ligand, and the abnormalities in brain development in CXCR4 knockout mice may be related to a lack of compensatory mechanisms. It may be necessary to combine knockouts of several chemokine receptors in order to tease out the function of this class of signal molecules.

6.3.2 Exploring the Detrimental Effects of Neuroinflammation on Neural Progenitor Cells: Unexpected Interactions and Novel Pathways

Recruitment of progenitor cells to regions of inflammation may have positive consequences in terms of repair, but because progenitor cells express cytokine receptors, they may become the unintended targets of a damaging immune response during inflammatory injury to the brain. Thus, immune factors that enhance the vulnerability of endogenous or transplanted progenitor cells would prevent effective repair; this may be an important part of disease progression in neurodegenerative diseases (Armstrong and Barker 2001).

Table 1. Receptor and ligands that may be part of the NPC-inflammatory interface (*continued on next page*)

Receptor/ligand	Some functional aspects in inflammation and hematopoiesis	Possible functional aspects in NPC biology
CCR2/ (Ji et al. 2004a) various CKs	Recruitment of inflammatory cells towards sites of inflammation (Kurihara et al. 1997; Daly and Rollins 2003; Babcock et al. 2003)	Migration?
CCR3/ (Krathwohl and Kaiser 2004) various CKs	HPC and Immune effector cell migration into inflamed tissue (Sehmi et al. 2003; Sabroe and Williams 2001) Maturation of HPCs in inflamed tissue (Lamkhioued et al. 2003)	Migration? (Krathwohl and Kaiser 2004)
CCR5/ (Ji et al. 2004a) various CKs	Immune effector cell recruitment to sites of inflammation (Wong and Fish 2003; Glabinski et al. 2002) T-cell maturation (Luther and Cyster 2001)	Migration? (Boutet et al. 2001) Integration? (Boutet et al. 2001)
CXCR4/ (Ji et al. 2004a) SDF1α	Extravasation (Peled et al. 1999) HSC homing (Peled et al. 1999; Ceradini et al. 2004)	Migration (Ehtesham et al. 2004) Extravasation (Allport and Shinde Patil 2004)
CXC3CR1/ (Ji et al. 2004a) Fractalkine	Extravasation of leukocytes (Fong et al. 1998)	Survival of NPCs (Krathwohl and Kaiser 2004) Migration? (Ji et al. 2004b)
LIF/gp130 (Wright et al. 2003)	Microglia proliferation and activation (Kerr and Patterson 2004) Macrophage chemotaxis (Sugiura et al. 2000) Anti-inflammatory functions (Gadient and Patterson 1999) Maturation of hematopoietic lineages (Metcalf 1997)	Self-renewal (Carpenter et al. 1999) Proliferation (Carpenter et al. 1999; Satoh and Yoshida 1997; Bauer et al. 2003) Neuronal differentiation (Potter and Ling 1999; Galli et al. 2000) Astrocytic differentiation (Guo et al. 1997)

Table 1. (*continued*)

Receptor/ligand	Some functional aspects in inflammation and hematopoiesis	Possible functional aspects in NPC biology
IL-1 Receptor/ ligand system (Dame et al. 2002)	Complex functions in the initiation and maintenance of inflammation and the immune response in the brain (Van der Meide and Schellekens 1996; Dinarello 1994; Boutin et al. 2003) Important factor for HPCs (Ruscetti 1994)	Neuronal differentiation (Potter and Ling 1999) Oligodendrocytic differentiation? (Vela et al. 2002)
Il-5 (Mehler et al. 1995)	Maturation of Eosinophils, B- and T-cells (Caldenhoven et al. 1998; Takatsu 1998; Apostolopoulos et al. 2000) Proliferation of microglia (Ringheim 1995) Downregulation of inflammation (Ziemssen et al. 2002)	Neuronal differentiation (Mehler et al. 1993)
IL-6/ (Monje and Toda 2003) gp130	Complex roles in regulation of inflammation (Munoz-Fernandez and Fresno 1998) Important factor in hematopoiesis (Ruscetti 1994)	Reduced hippocampal neurogenesis (Vallieres et al. 2002) Inhibition of neuronal differentiation (Monje and Toda 2003) Astrocytic differentiation (Nakashima and Taga 2002) Oligodendrocytic differentiation (Zhang et al. 2004)
IL-7	Maturation of B and T-cells	Promotes survival of progenitors (Rozental et al. 1995) Neuronal differentiation (Rozental et al. 1995; Michaelson et al. 1996)

Table 1. (*continued*)

Receptor/ligand	Some functional aspects in inflammation and hematopoiesis	Possible functional aspects in NPC biology
IL-9 (Mehler et al. 1995)	Proliferation and maturation of T-, B-mast cells and HPCs (Demoulin and Renauld 1998; Appasamy 1999)	Neuronal differentiation (Mehler et al. 1993)
IL-11/ gp130	Important role for maturation of several lineages in hematopoiesis (Kobayashi et al. 1994; Schwertschlag et al. 1999)	Neuronal differentiation (Mehler et al. 1993)
B7/CD28	Positive and negative immunostimulation (Carreno and Collins 2002)	NPC-immunological synapse, Induction of apoptosis (Imitola et al. 2004)
TNFα	Key player in inflammation (Stoll et al. 2002; Lenzlinger et al. 2001) Proregenerative? (Feuerstein et al. 1997)	Apoptosis (Imitola et al. 2004; Wong et al. 2004) Migration (Wu et al. 2000; Ben-Hur et al. 2003)
CD95/ Fas-ligand (Klassen et al. 2001)	Activation/Control/Termination of the immune response (Pender and Rist 2001; Pinkoski and Green 2000; Choi and Benveniste 2004)	Apoptosis? (Cheema et al. 2004) One substrate of NPCs immune privilege? (Ceccatelli et al. 2004)
c-kit/ SCF (Sun et al. 2004)	Mast cell and eosinophil activation (Lukacs et al. 1996) Complex role in the regulation of inflammation (Fantuzzi and Dinarello 1998; Zhang and Fedoroff 1998) Proliferation of myeloid progenitor in sites of inflammation	Proliferation (Jin et al. 2002) Migration (Sun et al. 2004)

We have investigated the possibility that neural progenitor cells may have interactions with immune cells such as T cells, asking whether they express molecules that may render this interaction functional. For example, we examined expression of costimulatory molecules such as CD80 and CD86, which deliver signals to T cells via the CD28 receptor, mediating increased cytokine production, proliferation, and cell viability (Hathcock et al. 1993) We demonstrated costimulatory molecule expression by neural progenitor cells and also observed that T cells have functional interactions with adult and neonatal neural progenitor cells (see Fig. 4; Imitola et al. 2004) We also reported on the differential regulation of costimulatory molecule expression by inflammatory or stress- inducing stimuli in vitro. Most interestingly, we showed that nestin-positive cells expressing CD80, which were presumably neural progenitor cells, increased during the acute phase of EAE. Others stem cells, such as mouse embryonic stem cells, have been reported to express CD80 and Class I MHCs (Ling et al. 1998). Others have found that neural progenitor cells express nonclassical costimulatory molecules such as the tetraspanin CD9 and CD81, although the function of such molecules on neural progenitor cells is not known. Taken together, these results suggest that neural progenitor cells in vivo may have interaction with immune cells via interaction with cell surface receptors.

How could neuroinflammation support neural progenitor cell-based brain repair?

Microglia, being of hematopoietic cell lineage, are early responders to brain damage such as hypoxic-ischemic brain injury. Microglia are able to change from their "resting" ramified state to an "activated" ameboid form, in order to migrate to areas of damage, proliferate and engulf invading organisms, apoptotic cells or cell debris (Andersson et al. 1991a,b et al.). They also secrete a variety of cytokines that create an inflammatory reaction that has both helpful and harmful effects on the neurons in the vicinity (Nakamura 2002; Popovich et al. 2002; Rogove et al. 2002; Stoll et al. 2002). Astrocytes also respond to the damage, either directly or indirectly by being stimulated by microglial inflammatory signals, but the interaction between these cells is poorly understood (Balasingam et al. 1996; Glabinski et al. 1996; Aldskogius and Kozlova 1998; Stoll et al. 1998; Hailer et al. 2001).

It has been suggested that microglia are kept in a resting state by intimate communication between neurons and microglia (Polazzi and Contestabile 2002). Some aspects of the microglial response to injury could be interpreted as aimed at neuroprotection, underpinned by the fact that the most prominent cytokines (Il-1ß, IL-6, TNF) in the inflammatory reaction of microglia have been shown not only to exert

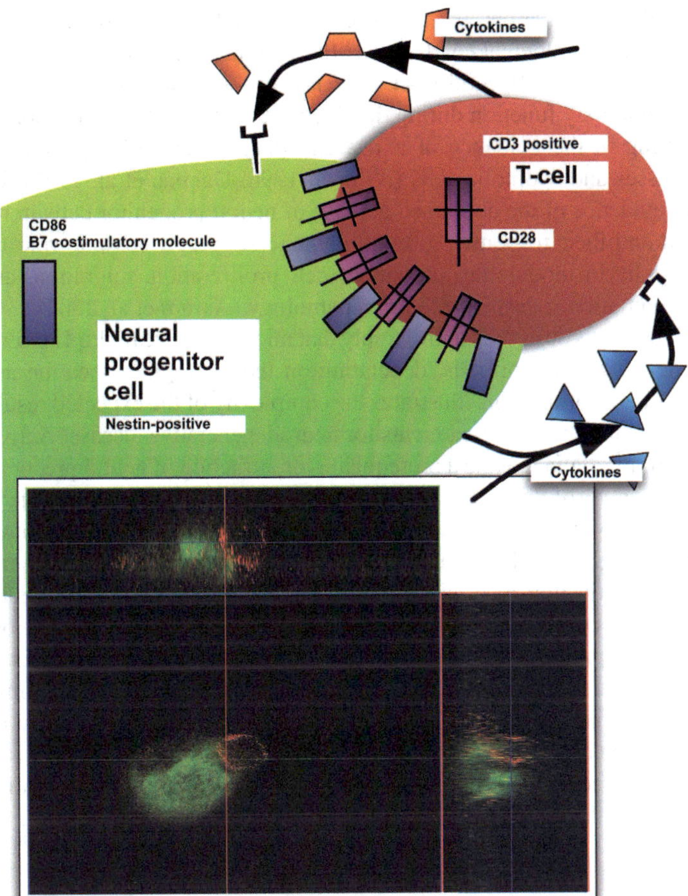

Fig. 4. A neural progenitor cell forms an immunological synapse with a T-cell (Imitola et al. 2004). Confocal phase microscopy shows a primary NSC-T cell conjugate and CD3 redistribution to the contact zone. NSC is stained green with nestin antibody and CD3 stained in red. Three dimensional reconstruction demonstrating the interaction between NSCs and T cells showing redistribution of CD3 to the contact zone. Schematic interpretation of this picture above the microscopic images: The NSC-T-cell complex is formed by interaction of costimulatory molecules (CD28 and CD80) and serves the focused interaction of single NPCs and T-cells

neurodegenerative but also neuroprotective and regenerative functions in certain pathological conditions (Munoz-Fernandez and Fresno 1998; Carlson et al. 1999; Stoll et al. 2002). It has also been suggested that microglia may function during development to remodel brain tissue by assisting in the guidance of axons and phagocytosis of degenerating processes and apoptotic cells (Ashwell 1990; Dalmau et al. 1997) Recent evidence of the continued activity of neural progenitor cells in the brain and their responses to brain damage suggest a possible role for microglia in modulating progenitor cell proliferation, migration, and differentiation in response to brain pathology (Aarum et al. 2003).

The results described above imply that microglia have multiple effects on neural progenitor cells, depending on the microglia's location and state of activation. This illustrates the complexity of the expected results of anti- inflammatory treatments for neurodegenerative disease. Activation of microglia may be a prerequisite for attraction of neural precursor cells to areas of damage. However, reports suggest that neural progenitor cell survival is enhanced by the administration of anti-inflammatory and microglial-inhibiting agents (Monje et al. 2003), indicating that microglial activation also appears to have detrimental effects on neural progenitor cells. Thus inflammation may act as a double-edged sword for neural progenitor cells; anti-inflammatory treatments may aid survival of neural progenitor cells, but, if administered at the wrong time, anti- inflammatory agents may work at cross purposes to repair by inhibiting the homing of progenitor cells. Understanding the intricacies of progenitor cell- microglial interactions will be pivotal for orchestrating treatments that are synergistic rather than antagonistic.

6.3.3 Gliomatropism of Neural Progenitor Cells: Cues from Regional Inflammatory Signals?

The interface of inflammatory processes with neural progenitor cells is likely to be of great importance in brain malignancy. The neural progenitor cells homing reaction to tumors seems to be similar to that in stroke lesions, but the functional relevance is not understood (Noble 2000; Staflin et al. 2004). Most research in this field is currently aimed at using the homing abilities of neural progenitor cells as a therapeutic approach for gene or drug delivery to the widespread and invasive pathology of primary or secondary brain tumors.

Brain tumors are invariably accompanied by a certain degree of regional inflammation (Davies 2002). Intracerebral trafficking of circulating inflammatory cells and lymphocytes occurs in the absence of an intact blood–brain barrier. Microglia also accumulate within the tumor, attracted by chemotactic factors produced by the tumor and its surrounding microenvironment (Badie et al. 1999; Platten et al. 2003). The repertoire of tumor-associated inflammatory cells and lymphocytes is dependent on the type of tumor as well as the tumor grade, likely in association with the degree local necrosis (Stevens et al. 1988; Badie and Schartner 2000; Parney et al. 2000; Walker et al. 2003).

The role of inflammatory responses in intracranial neoplasms is not well understood, but it is reasonable to assume that a certain degree of inflammation in the form of nonspecific as well as specific immunity is likely to be beneficial in control of brain tumors. However, neuroinflammation may perversely aid in brain tumor infiltration and dissemination (Weber and Ashkar 2000); it has been reported that tumor-associated macrophages (TAMs) recruited by monocyte chemoattractant protein-1 (MCP1) promoted the growth of glioma in vivo (Platten et al. 2003).

Neural progenitor cells have been shown to migrate over long distances to implanted glioma tumors in experimental animal models (Aboody et al. 2000; Benedetti et al. 2000). The mechanism is this apparent gliomatropism is not well understood, but researchers have suggested that neural progenitor cells follow molecular "breadcrumbs", acting as chemotactic beacons that are left behind by the tumor cells (Yip et al. 2003, 2004). Some of these signals include well-known inflammatory signals such as chemokines and adhesion molecules (Merzak et al. 1994; Akiyama et al. 2001). Recent experimental findings have also implicated a role for SDF1 and its receptor CXCR4 in mediating the migration of stem cells toward regions of CNS pathology. A recent report showed overexpression of SDF1 in regions of cerebral ischemia, which preferentially attracted the migration of bone marrow-derived inflammatory cells (Hill et al. 2004). SDF1 and CXCR4 are expressed in gliomas and their expression is proportionally increased with increasing histological grade (Rempel et al. 2000). Additional evidence for SDF1/CXCR4 involvement is provided by the report that inhibition of

CXCR4 inhibits neural progenitor cell gliomatropism in a rodent model (Ehtesham et al. 2004).

All of the neural precursor homing experiments described above were done with exogenous neural progenitors. It is still unclear whether endogenous neural progenitors residing in the brain are activated to migrate to sites of injury in all above mentioned brain pathologies. There is an additional intriguing possibility suggested by some researchers, that there is an endogenous pool of neural stem cells circulating in the blood (Pituch- Noworolska et al. 2003; Kucia et al. 2004). The evidence for this idea is circumstantial but interesting. Neural stem cells express the same integrins that are involved in the SDF1-mediated extravasation of leukocytes from the blood into inflamed peripheral tissues (Allport et al. 2004), suggesting that they may be capable of similar behavior. There has also been some success with intravascular delivery of neural progenitors; cells delivered in this way have been reported to migrate into the brain parenchyma if the blood–brain barrier is broken down (Pluchino et al. 2003).

6.4 Creating a Permissive Environment for Neural Progenitor Cells

Since the hippocampus and the subventricular zone appear to have the ability to generate new neurons, it seems likely that these regions maintain physiological conditions that are different from the majority of the brain that is not neurogenic (Eriksson et al. 1998; Gage 2002). In order to understand how we might intervene to encourage new neuron development, it is important that we examine both the permissive qualities of neurogenic environments, and the conditions that repress neurogenesis in most of the CNS.

In 1928 the Nobel laureate Ramon y Cajal proposed in his book on the degeneration and regeneration of the nervous system: "*In adult centres the nerve paths are something fixed, ended, immutable. Everything may die, nothing may be regenerated. It is for the science of the future to change, if possible, this harsh decree.*" (Cajal 1928)

In recent years we have been edging closer toward fulfilling Cajal's dream of neural regeneration, but there are many barriers still confronting us. We know that there is a population of neuroepithelial cells within the brain that can be stimulated to divide ex vivo and differentiate into neurons, astroglia, and oligodendrocytes. There might be some degree of normal repair in response to injury, but it is insufficient and the harsh reality is that no brain lesion in any animal experiment or any human has been fully "restored", either with endogenous or exogenous neural progenitor cells (Arvidsson et al. 2002).

There are many different kinds of neurodegeneration, and the approaches to regeneration will have to be tailored to the pathology. For example, it is an enormous challenge in cases in which tissue has been completely destroyed, such as in the necrotic region of a stroke, or in spinal cord injury, where the nervous system would have to be rebuilt from scratch. In contrast, there are situations in which it may be feasible to repair partially functional tissue, such as the penumbra in stroke, the diseased cortex in Alzheimer's disease, or the substantia nigra in Parkinson's disease. In each case, we must consider the limitations imposed on neural precursor cells by a hostile environment.

Many researchers are currently exploring methods of tissue engineering as a way of repairing the extensive lesions in stroke and spinal cord injury (Park et al. 2002; Teng et al. 2002). For partial lesions, a focus on the precursors themselves is proving to be more fruitful. Neural precursors are most active during embryological development, and the fetal brain appears to provide an environment that is supportive to integration of exogenous precursors. For example, neural progenitors transplanted in utero into a primate have been shown to participate in brain development, taking up residence in the well-described neurogenic niches of the hippocampus and subventricular zone. But they also spread all across the brain and remain in a undifferentiated state (Ourednik et al. 2001). This observation may give clues about the postdevelopmental fate of endogenous neural progenitor cells, which assume an astroglia-like phenotype and can be extracted and propagated from several regions of the adult human brain (Roy et al. 2000; Jain et al. 2003; Nunes et al. 2003; Schwartz et al. 2003; Sanai et al. 2004; Zhang, Klueber et al. 2004).

These observations suggest that in the adult brain there are factors that actively suppress the neurogenic potential of the quiescent exogenous neural stem cells (Palmer et al. 1999; Yamamoto et al. 2001). There is some evidence that suggests that the local astrocyte population may influence the neurogenic character of the environment. Astrocytes are not present during the most prominent phase of neurogenesis during brain development. Also, a recent study suggests regional differences in astrocyte's effects on neural progenitor cells. Astrocytes from the hippocampus were shown to be able to support proliferation of neural progenitor cells, whereas astrocytes extracted from the spinal cord inhibited proliferation of progenitor cells (Song et al. 2002). This observation suggests some interesting possibilities for a role of astrocytes in neurogenesis. For example, one could ask whether the role of radial glia cells/astrocytes is similar or different in neurogenesis that takes place during development and that occurring in the adult in response to injury. If regionally specific astrocytes affect neurogenesis differently, would it be possible to reproduce favorable neurogenic conditions by therapeutically changing the character of astrocytes?

6.5 Summary

Neural stem and progenitor cells express a variety of receptors that enable them to sense and react to signals emanating from physiological and pathophysiological conditions in the brain as well as elsewhere in the body. Many of these receptors and were first described in investigations of the immune system, particularly with respect to hematopoietic stem cells. This emerging view of neurobiology has two major implications. First, many phenomena known from the hematopoietic system may actually be generalizable to stem cells from many organ systems, reflecting the cells' progenitor-mediated regenerative potential. Second, regenerative interfaces may exist between diverse organ systems; populations of cells of neuroectodermal and hematopoietic origin may interact to play a crucial role in normal brain physiology, pathology, and repair. An understanding of the origins of signals and the neural progenitors' responses might lead to the development of effective therapeutic strategies to counterbalance acute and chronic neurodegenerative processes. Such strategies may include modifying and modulating cells with regenerative potential in subtle ways. For example, stem cells might be able to detect pathology-associated signals and be used as "interpreters" to mediate drug and other therapeutic interventions.

This review has focused on the role of inflammation in brain repair. We propose that resident astroglia and blood-born cells both contribute to an inflammatory signature that is unique to each kind of neuronal degeneration or injury. These cells play a key role in coordinating the neural progenitor cell response to brain injury by exerting direct and indirect environmentally mediated influence on neural progenitor cells.

We suggest that investigations of the neural progenitor -immunologic interface will provide valuable data related to the mechanisms by which endogenous and exogenous neural progenitor cells react to brain pathology, ultimately aiding in the design of more effective therapeutic applications of stem cell biology. Such improvements will include: (1) ascertaining the proper timing for implanting exogenous neural progenitor cells in relation to the administration of anti-inflammatory agents; (2) identifying what types of molecules might be administered during injury to enhance the mobilization and differentiation of endogenous and exogenous neural progenitor cells while also inhibiting the detrimental aspects of the inflammatory reaction; (3) divining clues as to which molecules may be required to change the lesioned environment in order to invite the homing of reparative neural progenitor cells.

References

Aarum J, Sandberg K, et al. (2003) Migration and differentiation of neural precursor cells can be directed by microglia. Proc Natl Acad Sci 100(26):15983–15988

Aboody KS, Brown A, et al. (2000) Neural stem cells display extensive tropism for pathology in adult brain: evidence from intracranial gliomas. Proc Natl Acad Sci 97(23):12846–12851

Akiyama Y, Jung S, et al. (2001) Hyaluronate receptors mediating glioma cell migration and proliferation. J Neurooncol 53(2):115–127

Aldskogius H, Kozlova EN (1998) Central neuron-glial and glial-glial interactions following axon injury. Prog Neurobiol 55(1):1–26

Allport JR, Shinde Patil VR, Weissleder R (2004) Murine neuronal progenitor cells are preferentially recruited to tumor vasculature via alpha (4)-integrin and SDF-1alpha-dependent mechanisms. Cancer Biol Ther 3(9)

Alvarez-Buylla A, Lois C (1995) Neuronal stem cells in the brain of adult vertebrates. Stem Cells 13(3):263–272

Andersson PB, Perry VH, et al. (1991a) The CNS acute inflammatory response to excitotoxic neuronal cell death. Immunol Lett 30(2):177–181

Andersson PB, Perry VH, et al. (1991b) The kinetics and morphological characteristics of the macrophage-microglial response to kainic acid-induced neuronal degeneration. Neuroscience 42(1):201–214

Apostolopoulos V, et al. (2000) A role for IL-5 in the induction of cytotoxic T lymphocytes in vivo. Eur J Immunol 30(6):1733–1739

Appasamy PM (1999) Biological and clinical implications of interleukin-7 and lymphopoiesis. Cytokines Cell Mol Ther 5(1):25–39

Armstrong RJ, Barker RA (2001) Neurodegeneration: a failure of neuroregeneration? Lancet 358(9288):1174–1176

Arvidsson A, et al. (2002) Neuronal replacement from endogenous precursors in the adult brain after stroke. Nat Med 8(9):963–970

Ashwell K (1990) Microglia and cell death in the developing mouse cerebellum. Brain Res Dev Brain Res 55(2):219–230

Babcock AA, et al. (2003) Chemokine expression by glial cells directs leukocytes to sites of axonal injury in the CNS. J Neurosci 23(21):7922–7930

Badie B, et al. (1999) In vitro modulation of microglia motility by glioma cells is mediated by hepatocyte growth factor/scatter factor. Neurosurgery 44(5):1077–1082; discussion 1082–1083

Badie B, Schartner JM (2000) Flow cytometric characterization of tumor-associated macrophages in experimental gliomas. Neurosurgery 46(4):957–961; discussion 961–962

Bagri A, Gurney T, et al. (2002) The chemokine SDF1 regulates migration of dentate granule cells. Development 129(18):4249–4260

Baker CA, Manuelidis L (2003) Unique inflammatory RNA profiles of microglia in Creutzfeldt-Jakob disease. Proc Natl Acad Sci 100(2):675–679

Balasingam V, Dickson K, et al. (1996) Astrocyte reactivity in neonatal mice: apparent dependence on the presence of reactive microglia/macrophages. Glia 18(1):11–26

Bauer S, et al. (2003) Leukemia inhibitory factor is a key signal for injury-induced neurogenesis in the adult mouse olfactory epithelium. J Neurosci 23(5):1792–1803

Benedetti S, Pirola B, et al. (2000) Gene therapy of experimental brain tumors using neural progenitor. Nat Med 6(4):447–450

Ben-Hur T, Ben-Menachem O, et al. (2003) Effects of proinflammatory cytokines on the growth, fate, and motility of multipotential neural precursor cells. Mol Cell Neurosci 24(3):623–631

Boutet A, et al. (2001) Cellular expression of functional chemokine receptor CCR5 and CXCR4 in human embryonic neurons. Neurosci Lett 311(2):105–108

Boutin H, et al. (2003) The expanding interleukin-1 family and its receptors: do alternative IL-1 receptor/signaling pathways exist in the brain? Mol Neurobiol 27(3):239–248

Brett FM, Mizisin AP, et al. (1995) Evolution of neuropathologic abnormalities associated with blood-brain barrier breakdown in transgenic mice expressing interleukin-6 in astrocytes. J Neuropathol Exp Neurol 54(6):766–775

Brionne TC, Tesseur I, et al. (2003) Loss of TGF-beta 1 leads to increased neuronal cell death and microgliosis in mouse brain. Neuron 40(6):1133–1145

Cajal SR (1928) Degeneration and regeneration of the nervous system. Oxford University Press, Oxford

Caldenhoven E, et al. (1998) Differential activation of functionally distinct STAT5 proteins by IL-5 and GM-CSF during eosinophil and neutrophil differentiation from human CD34+ hematopoietic stem cells. Stem Cells 16(6):397–403

Campbell IL, Abraham CR, et al. (1993) Neurologic disease induced in transgenic mice by cerebral overexpression of interleukin 6. Proc Natl Acad Sci 90(21):10061–10065

Carlson NG, Wieggel WA, et al. (1999) Inflammatory cytokines IL-1 alpha, IL-1 beta, IL-6, and TNF-alpha impart neuroprotection to an excitotoxin through distinct pathways. J Immunol 163(7):3963–3968

Carpenter MK, et al. (1999) In vitro expansion of a multipotent population of human neural progenitor cells. Exp Neurol 158(2):265–278

Carreno BM, Collins M (2002) The B7 family of ligands and its receptors: new pathways for costimulation and inhibition of immune responses. Annu Rev Immunol 20:29–53

Ceccatelli S, et al. (2004) Neural stem cells and cell death. Toxicol Lett 149(1–3):59–66

Ceradini DJ, et al. (2004) Progenitor cell trafficking is regulated by hypoxic gradients through HIF-1 induction of SDF-1. Nat Med 10(8):858–864

Cheema ZF, et al. (2004) The extracellular matrix, p53 and estrogen compete to regulate cell-surface Fas/Apo-1 suicide receptor expression in proliferating embryonic cerebral cortical precursors, and reciprocally, Fas-ligand modifies estrogen control of cell-cycle proteins. BMC Neurosci 5(1):11

Chiang CS, Stalder A, et al. (1994) Reactive gliosis as a consequence of interleukin-6 expression in the brain: studies in transgenic mice. Dev Neurosci 16(3–4):212–221

Choi C, Benveniste EN (2004) Fas ligand/Fas system in the brain: regulator of immune and apoptotic responses. Brain Res Brain Res Rev 44(1):65–81

Dalmau I, Finsen B, et al. (1997) Development of microglia in the prenatal rat hippocampus. J Comp Neurol 377(1):70–84

Daly C, Rollins BJ (2003) Monocyte chemoattractant protein-1 (CCL2) in inflammatory disease and adaptive immunity: therapeutic opportunities and controversies. Microcirculation 10(3–4):247–257

Dame JB, et al. (2002) The effect of interleukin-1beta (IL-1beta) and tumor necrosis factor-alpha (TNF-alpha) on granulocyte macrophage-colony stimulating factor (GM-CSF) production by neuronal precursor cells. Eur Cytokine Netw 13(1):128–133

Davies DC (2002) Blood-brain barrier breakdown in septic encephalopathy and brain tumours. J Anat 200(6):639–646

Demoulin JB, Renauld JC (1998) Interleukin 9 and its receptor: an overview of structure and function. Int Rev Immunol 16(3–4):345–364

Dinarello CA (1994) The biological properties of interleukin-1. Eur Cytokine Netw 5(6):517–531

Ehtesham M, et al. (2004) Glioma tropic neural stem cells consist of astrocytic precursors and their migratory capacity is mediated by CXCR4. Neoplasia 6(3):287–293

Eriksson PS, Perfilieva E, et al. (1998) Neurogenesis in the adult human hippocampus. Nat Med 4(11):1313–1317

Eskes C, Juillerat-Jeanneret L, et al. (2003) Involvement of microglia- neuron interactions in the tumor necrosis factor-alpha release, microglial activation, and neurodegeneration induced by trimethyltin. J Neurosci Res 71(4):583–590

Fantuzzi G, Dinarello CA (1998) Stem cell factor-deficient mice have a dysregulation of cytokine production during local inflammation. Eur Cytokine Netw 9(1):85–92

Feuerstein GZ, Wang X, Barone FC (1997) Inflammatory gene expression in cerebral ischemia and trauma. Potential new therapeutic targets. Ann N Y Acad Sci 825:179–193

Flax JD, Aurora S, et al. (1998) Engraftable human neural stem cells respond to developmental cues, replace neurons, and express foreign genes. Nat Biotechnol 16(11):1033–1039

Fong AM, et al. (1998) Fractalkine and CX3CR1 mediate a novel mechanism of leukocyte capture, firm adhesion, and activation under physiologic flow. J Exp Med 188(8):1413–1419

Gadient RA, Patterson PH (1999) Leukemia inhibitory factor, Interleukin 6, and other cytokines using the GP130 transducing receptor: roles in inflammation and injury. Stem Cells 17(3):127–137

Gage FH (2002) Neurogenesis in the adult brain. J Neurosci 22(3):612–613

Galli R, et al. (2000) Regulation of neuronal differentiation in human CNS stem cell progeny by leukemia inhibitory factor. Dev Neurosci 22(1–2):86– 95

Glabinski AR, Balasingam V, et al. (1996) Chemokine monocyte chemoattractant protein-1 is expressed by astrocytes after mechanical injury to the brain. J Immunol 156(11):4363–4368

Glabinski AR, et al. (2002) Experimental autoimmune encephalomyelitis: CC chemokine receptor expression by trafficking cells. J Autoimmun 19(4):175–181

Graeber MB, Scheithauer BW, et al. (2002) Microglia in brain tumors. Glia 40(2):252–259

Guo X, et al. (1997) Leukemia inhibitory factor and ciliary neurotrophic factor regulate dendritic growth in cultures of rat sympathetic neurons. Brain Res Dev Brain Res 104(1–2):101–110

Hailer NP, Wirjatijasa F, et al. (2001) Astrocytic factors protect neuronal integrity and reduce microglial activation in an in vitro model of N- methyl-D-aspartate-induced excitotoxic injury in organotypic hippocampal slice cultures. Eur J Neurosci 14(2):315–326

Han HS, Yenari MA (2003) Cellular targets of brain inflammation in stroke. Curr Opin Investig Drugs 4(5):522–529

Hathcock KS, Laszlo G, et al. (1993) Identification of an alternative CTLA-4 ligand costimulatory for T cell activation. Science 262(5135):905–907

Hill JR, Kuriyama N, et al. (1999) Molecular genetics of brain tumors. Arch Neurol 56(4):439–441

Hill WD, Hess DC, et al. (2004) SDF-1 (CXCL12) is upregulated in the ischemic penumbra following stroke: association with bone marrow cell homing to injury. J Neuropathol Exp Neurol 63(1):84–96

Imitola J, Comabella M, et al. (2004) Neural stem/progenitor cells express costimulatory molecules that are differentially regulated by inflammatory and apoptotic stimuli. Am J Pathol 164(5):1615–1625

Jain M, Armstrong RJ, et al. (2003) GABAergic immunoreactivity is predominant in neurons derived from expanded human neural precursor cells in vitro. Exp Neurol 182(1):113–123

Ji JF, et al. (2004a) Expression of chemokine receptors CXCR4, CCR2, CCR5 and CX3CR1 in neural progenitor cells isolated from the subventricular zone of the adult rat brain. Neurosci Lett 355(3):236–240

Ji JF, et al. (2004b) Interactions of chemokines and chemokine receptors mediate the migration of mesenchymal stem cells to the impaired site in the brain after hypoglossal nerve injury. Stem Cells 22(3):415–427

Jin K, et al. (2002) Stem cell factor stimulates neurogenesis in vitro and in vivo. J Clin Invest 110(3):311–319

Jin K, Sun Y, et al. (2003) Directed migration of neuronal precursors into the ischemic cerebral cortex and striatum. Mol Cell Neurosci 24(1):171–189

Kerr BJ, Patterson PH (2004) Potent pro-inflammatory actions of leukemia inhibitory factor in the spinal cord of the adult mouse. Exp Neurol 188(2):391–407

Kielian T (2004) Microglia and chemokines in infectious diseases of the nervous system: views and reviews. Front Biosci 9:732–750

Klassen H, et al. (2001) Surface markers expressed by multipotent human and mouse neural progenitor cells include tetraspanins and nonprotein epitopes. Neurosci Lett 312(3):180–182

Kobayashi S, et al. (1994) Interleukin-11. Leuk Lymphoma 15(1–2):45–49

Krathwohl MD, Kaiser JL (2004) Chemokines promote quiescence and survival of human neural progenitor cells. Stem Cells 22(1):109–118

Kucia M, Ratajczak J, et al. (2004) Tissue-specific muscle, neural and liver stem/progenitor cells reside in the bone marrow, respond to an SDF-1 gradient and are mobilized into peripheral blood during stress and tissue injury. Blood Cells Mol Dis 32(1):52–57

Kurihara T, et al. (1997) Defects in macrophage recruitment and host defense in mice lacking the CCR2 chemokine receptor. J Exp Med 186(10):1757–1762

Lamkhioued B, et al. (2003) The CCR3 receptor is involved in eosinophil differentiation and is up-regulated by Th2 cytokines in CD34+ progenitor cells. J Immunol 170(1):537–547

Lazarini F, Tham TN, et al. (2003) Role of the alpha-chemokine stromal cell-derived factor (SDF-1) in the developing and mature central nervous system. Glia 42(2):139–148

Lenzlinger PM, et al. (2001) The duality of the inflammatory response to traumatic brain injury. Mol Neurobiol 24(1–3):169–181

Liberto CM, Albrecht PJ, et al. (2004) Pro-regenerative properties of cytokine-activated astrocytes. J Neurochem 89(5):1092–1100

Ling V, Munroe RC, et al. (1998) Embryonic stem cells and embryoid bodies express lymphocyte costimulatory molecules. Exp Cell Res 241(1):55–65

Loddick SA, Turnbull AV, et al. (1998) Cerebral interleukin-6 is neuroprotective during permanent focal cerebral ischemia in the rat. J Cereb Blood Flow Metab 18(2):176–179

Lu M, Grove EA, et al. (2002) Abnormal development of the hippocampal dentate gyrus in mice lacking the CXCR4 chemokine receptor. Proc Natl Acad Sci 99(10):7090–7095

Lukacs NW, et al. (1996) The role of stem cell factor (c-kit ligand) and inflammatory cytokines in pulmonary mast cell activation. Blood 87(6):2262–2268

Luther SA, Cyster JG (2001) Chemokines as regulators of T cell differentiation. Nat Immunol 2(2):102–107

Lynch WP, Sharpe AH, et al. (1999) Neural stem cells as engraftable packaging lines can mediate gene delivery to microglia: evidence from studying retroviral env-related neurodegeneration. J Virol 73(8):6841–6851

Ma Q, Jones D, et al. (1998) Impaired B-lymphopoiesis, myelopoiesis, and derailed cerebellar neuron migration in CXCR4- and SDF-1-deficient mice. Proc Natl Acad Sci 95(16):9448–9453

McKay R (1997) Stem cells in the central nervous system. Science 276(5309):66–71

Mehler MF, et al. (1993) Cytokine regulation of neuronal differentiation of hippocampal progenitor cells. Nature 362(6415):62–65

Mehler MF, et al. (1995) Cytokines regulate the cellular phenotype of developing neural lineage species. Int J Dev Neurosci 13(3–4):213–240

Meng XL, Shen JS, et al. (2003) Brain transplantation of genetically engineered human neural stem cells globally corrects brain lesions in the mucopolysaccharidosis type VII mouse. J Neurosci Res 74(2):266–277

Merzak A, Koocheckpour S, et al. (1994) CD44 mediates human glioma cell adhesion and invasion in vitro. Cancer Res 54(15):3988–3992

Metcalf D (1997) Murine hematopoietic stem cells committed to macrophage/dendritic cell formation: stimulation by Flk2-ligand with enhancement by regulators using the gp130 receptor chain. Proc Natl Acad Sci 94(21):11552–11556

Michaelson MD, et al. (1996) Interleukin-7 is trophic for embryonic neurons and is expressed in developing brain. Dev Biol 179(1):251–263

Modo M, Stroemer RP, et al. (2002) Effects of implantation site of stem cell grafts on behavioral recovery from stroke damage. Stroke 33(9):2270–2278

Modo M, Mellodew K, et al. (2004) Mapping transplanted stem cell migration after a stroke: a serial, in vivo magnetic resonance imaging study. Neuroimage 21(1):311–317

Monje ML, Toda H, et al. (2003) Inflammatory blockade restores adult hippocampal neurogenesis. Science 302(5651):1760–1765

Munoz-Fernandez MA, Fresno M (1998) The role of tumour necrosis factor, interleukin 6, interferon-gamma and inducible nitric oxide synthase in the development and pathology of the nervous system. Prog Neurobiol 56(3):307–340

Nakamura Y (2002) Regulating factors for microglial activation. Biol Pharm Bull 25(8):945–953

Nakashima K, Taga T (2002) Mechanisms underlying cytokine- mediated cell-fate regulation in the nervous system. Mol Neurobiol 25(3):233–244

Noble M (2000) Can neural stem cells be used to track down and destroy migratory brain tumor cells while also providing a means of repairing tumor-associated damage? Proc Natl Acad Sci 97(23):12393–12395

Nunes MC, Roy NS, et al. (2003) Identification and isolation of multipotential neural progenitor cells from the subcortical white matter of the adult human brain. Nat Med 9(4):439–447

Ourednik V, Ourednik J, et al. (2001) Segregation of human neural stem cells in the developing primate forebrain. Science 293(5536):1820–1824

Ourednik J, Ourednik V, et al. (2002) Neural stem cells display an inherent mechanism for rescuing dysfunctional neurons. Nat Biotechnol 20(11):1103–1110

Palmer TD, Takahashi J, et al. (1997) The adult rat hippocampus contains primordial neural stem cells. Mol Cell Neurosci 8(6):389–404

Palmer TD, Markakis EA, et al. (1999) Fibroblast growth factor-2 activates a latent neurogenic program in neural stem cells from diverse regions of the adult CNS. J Neurosci 19(19):8487–8497

Park KI, Teng YD, et al. (2002) The injured brain interacts reciprocally with neural stem cells supported by scaffolds to reconstitute lost tissue. Nat Biotechnol 20(11):1111–1117

Parney IF, Hao C, et al. (2000) Glioma immunology and immunotherapy. Neurosurgery 46(4):778–791; discussion 791–792

Peled A, et al. (1999) Dependence of human stem cell engraftment and repopulation of NOD/SCID mice on CXCR4. Science 283(5403):845–848

Pender MP, Rist MJ (2001) Apoptosis of inflammatory cells in immune control of the nervous system: role of glia. Glia 36(2):137–144

Peng H, Huang Y, et al. (2004) Stromal cell-derived factor 1- mediated CXCR4 signaling in rat and human cortical neural progenitor cells. J Neurosci Res 76(1):35–50

Penkowa M, Moos T, et al. (1999) Strongly compromised inflammatory response to brain injury in interleukin-6-deficient mice. Glia 25(4):343–357

Penkowa M, Camats J, et al. (2003a) Astrocyte-targeted expression of interleukin-6 protects the central nervous system during neuroglial degeneration induced by 6-aminonicotinamide. J Neurosci Res 73(4):481–496

Penkowa M, Giralt M, et al. (2003b) Astrocyte-targeted expression of IL-6 protects the CNS against a focal brain injury. Exp Neurol 181(2):130–148

Picard-Riera N, Decker L, et al. (2002) Experimental autoimmune encephalomyelitis mobilizes neural progenitors from the subventricular zone to undergo oligodendrogenesis in adult mice. Proc Natl Acad Sci 99(20):13211–13216

Pinkoski MJ, Green DR (2000) Cloak and dagger in the avoidance of immune surveillance. Curr Opin Genet Dev 10(1):114-9

Pituch-Noworolska A, Majka M, et al. (2003) Circulating CXCR4- positive stem/progenitor cells compete for SDF-1-positive niches in bone marrow, muscle and neural tissues: an alternative hypothesis to stem cell plasticity. Folia Histochem Cytobiol 41(1):13–21

Platten M, Kretz A, et al. (2003) Monocyte chemoattractant protein-1 increases microglial infiltration and aggressiveness of gliomas. Ann Neurol 54(3):388–392

Pluchino S, Quattrini A, et al. (2003) Injection of adult neurospheres induces recovery in a chronic model of multiple sclerosis. Nature 422(6933):688–694

Polazzi E, Contestabile A (2002) Reciprocal interactions between microglia and neurons: from survival to neuropathology. Rev Neurosci 13(3):221–242

Polazzi E, Contestabile A (2003) Neuron-conditioned media differentially affect the survival of activated or unstimulated microglia: evidence for neuronal control on apoptotic elimination of activated microglia. J Neuropathol Exp Neurol 62(4):351–362

Popovich PG, Guan Z, et al. (2002) The neuropathological and behavioral consequences of intraspinal microglial/macrophage activation. J Neuropathol Exp Neurol 61(7):623–633

Potter ED, Ling ZD, Carvey PM (1999) Cytokine-induced conversion of mesencephalic-derived progenitor cells into dopamine neurons. Cell Tissue Res 296(2):235–246

Rempel SA, Dudas S, et al. (2000) Identification and localization of the cytokine SDF1 and its receptor, CXC chemokine receptor 4, to regions of necrosis and angiogenesis in human glioblastoma. Clin Cancer Res 6(1):102–111

Ringheim GE (1995) Mitogenic effects of interleukin-5 on microglia. Neurosci Lett 201(2):131–134

Rogove AD, Lu W, et al. (2002) Microglial activation and recruitment, but not proliferation, suffice to mediate neurodegeneration. Cell Death Differ 9(8):801–806

Roy NS, Wang S, et al. (2000) In vitro neurogenesis by progenitor cells isolated from the adult human hippocampus. Nat Med 6(3):271–277

Rozental R, et al. (1995) Differentiation of hippocampal progenitor cells in vitro: temporal expression of intercellular coupling and voltage- and ligand-gated responses. Dev Biol 167(1):350–362

Ruscetti FW (1994) Hematologic effects of interleukin-1 and interleukin-6. Curr Opin Hematol 1(3):210–215

Sabroe I, Williams TJ, Pease JE (2001) Roles of chemokines in the regulation of leukocyte recruitment. Clin Sci (Lond) 100(4):359–362

Sanai N, Tramontin AD, et al. (2004) Unique astrocyte ribbon in adult human brain contains neural stem cells but lacks chain migration. Nature 427(6976):740–744

Satoh M, Yoshida T (1997) Promotion of neurogenesis in mouse olfactory neuronal progenitor cells by leukemia inhibitory factor in vitro. Neurosci Lett 225(3):165–168

Schwartz PH, Bryant PJ, et al. (2003) Isolation and characterization of neural progenitor cells from postmortem human cortex. J Neurosci Res 74(6):838–851

Schwertschlag US, et al. (1999) Hematopoietic, immunomodulatory and epithelial effects of interleukin-11. Leukemia 13(9):1307–1315

Sehmi R, et al. (2003) Allergen-induced fluctuation in CC chemokine receptor 3 expression on bone marrow CD34+ cells from asthmatic subjects: significance for mobilization of haemopoietic progenitor cells in allergic inflammation. Immunology 109(4):536–546

Shafer LL, McNulty JA, et al. (2003) Brain activation of monocyte lineage cells: brain-derived soluble factors differentially regulate BV2 microglia and peripheral macrophage immune functions. Neuroimmunomodulation 10(5):283–294

Snyder EY, Taylor RM, et al. (1995) Neural progenitor cell engraftment corrects lysosomal storage throughout the MPS VII mouse brain. Nature 374(6520):367–370

Song H, Stevens CF, et al. (2002) Astroglia induce neurogenesis from adult neural stem cells. Nature 417(6884):39–44

Staflin K, Honeth G, et al. (2004) Neural progenitor cell lines inhibit rat tumor growth in vivo. Cancer Res 64(15):5347–5354

Stevens A, Kloter I, et al. (1988) Inflammatory infiltrates and natural killer cell presence in human brain tumors. Cancer 61(4):738–743

Stoll G, Jander S, et al. (1998) Inflammation and glial responses in ischemic brain lesions. Prog Neurobiol 56(2):149–171

Stoll G, Jander S, et al. (2002) Detrimental and beneficial effects of injury-induced inflammation and cytokine expression in the nervous system. Adv Exp Med Biol 513:87–113

Strik HM, Stoll M, et al. (2004) Immune cell infiltration of intrinsic and metastatic intracranial tumours. Anticancer Res 24(1):37–42

Stumm RK, Zhou C, et al. (2003) CXCR4 regulates interneuron migration in the developing neocortex. J Neurosci 23(12):5123–5130

Sugiura S, et al. (2000) Leukaemia inhibitory factor is required for normal inflammatory responses to injury in the peripheral and central nervous systems in vivo and is chemotactic for macrophages in vitro. Eur J Neurosci 12(2):457–466

Sun L, Lee J, Fine HA (2004) Neuronally expressed stem cell factor induces neural stem cell migration to areas of brain injury. J Clin Invest 113(9):1364–1374

Szpak GM, Lechowicz W, et al. (2001) Neurones and microglia in central nervous system immune response to degenerative processes. Part 1: Alzheimer's disease and Lewy body variant of Alzheimer's disease. Quantitative study. Folia Neuropathol 39(3):181–192

Takatsu K (1998) Interleukin 5 and B cell differentiation. Cytokine Growth Factor Rev 9(1):25–35

Tan KT, Lip GY, et al. (2003) Post-stroke inflammatory response: effects of stroke evolution and outcome. Curr Atheroscler Rep 5(4):245–251

Tang Y, Shah K, et al. (2003) In vivo tracking of neural progenitor cell migration to glioblastomas. Hum Gene Ther 14(13):1247–1254

Teng YD, Lavik EB, et al. (2002) Functional recovery following traumatic spinal cord injury mediated by a unique polymer scaffold seeded with neural stem cells. Proc Natl Acad Sci 99(5):3024–3029

Tran PB, Miller RJ (2003) Chemokine receptors: signposts to brain development and disease. Nat Rev Neurosci 4(6):444–455

Vallieres L, Campbell IL, et al. (2002) Reduced hippocampal neurogenesis in adult transgenic mice with chronic astrocytic production of interleukin-6. J Neurosci 22(2):486–492

Van der Meide PH, Schellekens H (1996) Cytokines and the immune response. Biotherapy 8(3–4):243–249

Vela JM, et al. (2002) Interleukin-1 regulates proliferation and differentiation of oligodendrocyte progenitor cells. Mol Cell Neurosci 20(3):489–502

Wada R, Tifft CJ, et al. (2000) Microglial activation precedes acute neurodegeneration in Sandhoff disease and is suppressed by bone marrow transplantation. Proc Natl Acad Sci 97(20):10954–10959

Walker PR, Calzascia T, et al. (2003) T-cell immune responses in the brain and their relevance for cerebral malignancies. Brain Res Brain Res Rev 42(2):97–122

Weber GF, Ashkar S (2000) Molecular mechanisms of tumor dissemination in primary and metastatic brain cancers. Brain Res Bull 53(4):421–424

Weiss S, Dunne C, et al. (1996) Multipotent CNS stem cells are present in the adult mammalian spinal cord and ventricular neuroaxis. J Neurosci 16(23):7599–7609

Wong MM, EN Fish (2003) Chemokines: attractive mediators of the immune response. Semin Immunol 15(1):5–14

Wong G, Goldshmit Y, Turnley AM (2004) Interferon-gamma but not TNF alpha promotes neuronal differentiation and neurite outgrowth of murine adult neural stem cells. Exp Neurol 187(1):171–177

Wright LS, et al. (2003) Gene expression in human neural stem cells: effects of leukemia inhibitory factor. J Neurochem 86(1):179–195

Wu JP, et al. (2000) Tumor necrosis factor-alpha modulates the proliferation of neural progenitors in the subventricular/ventricular zone of adult rat brain. Neurosci Lett 292(3):203–206

Yamamoto S, Nagao M, et al. (2001) Transcription factor expression and Notch-dependent regulation of neural progenitors in the adult rat spinal cord. J Neurosci 21(24):9814–9823

Yandava BD, Billinghurst LL, et al. (1999) Global cell replacement is feasible via neural stem cell transplantation: evidence from the dysmyelinated shiverer mouse brain. Proc Natl Acad Sci 96(12):7029–7034

Yip S, Aboody KS, et al. (2003) Neural stem cell biology may be well suited for therapies. Cancer J 9(3):189–204

Yip S, Sidman RL, et al. (2004) Stem cells for targeting CNS malignancy. Karger

Zhang P, et al. (2004a) Enhancement of oligodendrocyte differentiation from murine embryonic stem cells by an activator of gp130 signaling. Stem Cells 22(3):344–354

Zhang R, Zhang Z, et al. (2004b) Activated neural stem cells contribute to stroke-induced neurogenesis and neuroblast migration toward the infarct boundary in adult rats. J Cereb Blood Flow Metab 24(4):441–448

Zhang SC, Fedoroff S (1998) Modulation of microglia by stem cell factor. J Neurosci Res 53(1):29–37

Zhang X, Klueber KM, et al. (2004) Adult human olfactory neural progenitors cultured in defined medium. Exp Neurol 186(2):112–123

Ziemssen T, et al. (2002) Glatiramer acetate-specific T-helper 1- and 2-type cell lines produce BDNF: implications for multiple sclerosis therapy. Brain-derived neurotrophic factor. Brain 125 (Pt 11):2381–2391

Zou YR, Kottmann AH, et al. (1998) Function of the chemokine receptor CXCR4 in haematopoiesis and in cerebellar development. Nature 393(6685):595–599

7 Remyelination and Restoration of Axonal Function by Glial Cell Transplantation

I.D. Duncan

7.1 Introduction

Myelin is a ubiquitous organelle throughout the central nervous system (CNS). While it is predominantly found in the white matter, it is also present around select axons in the grey matter. Myelin is essential for the normal conduction of impulses throughout the CNS. If myelin is lost as a result of an acquired or inherited disease (demyelination), or fails to develop normally as a result of a mutation (hypomyelination or dysmyelination), then nerve fibers conduct slowly or impulse conduction is blocked. This results in neurologic dysfunction of a widespread nature depending upon the tracts of the CNS that are involved. In multiple sclerosis (MS), the most important myelin disorder, neurologic dysfunction commonly affects the optic nerve (optic neuritis) and the long motor and sensory tracts of the spinal cord leading to paresis and sensory disturbance in the limbs and trunk.

In MS, demyelination is usually followed by endogenous repair or remyelination in the early course of the disease (Brück et al. 2003; Lassmann et al. 1997). However, after a period of time, remyelination fails and areas of chronic myelin loss or demyelinated plaques develop. More recently, the fate of axons in MS have become a target of research as it has become apparent that axons in MS are lost as well as the myelin sheath, both early in disease but more significantly, as time progresses (Bitsch et al. 2000; Ferguson et al. 1997; Kornek et al. 1999; Bjartmar et al. 2001; Trapp et al. 1998). It is thought that axonal loss is the key to long term disability; hence preventing this from occurring has become a major therapeutic goal. While neuroprotection has risen to the forefront of MS research, the simplest approach to protecting axons might be to simply restore the lost myelin by supplying myelinating cells from exogenous sources by transplantation, or by promoting endogenous repair. It is known that cells of the oligodendrocyte lineage may persist in

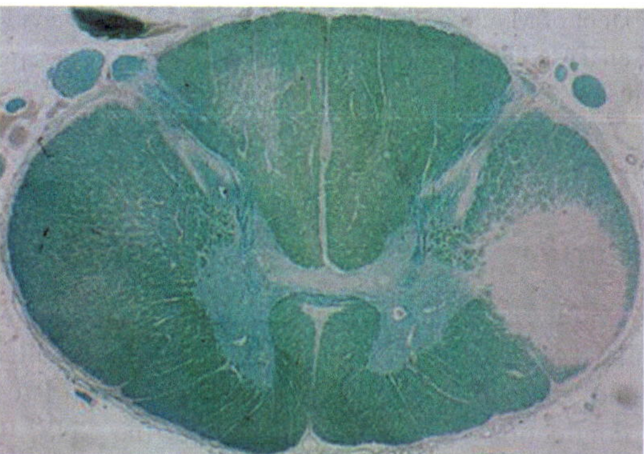

Fig. 1. A large plaque in the lateral column of the spinal cord of a patient with chronic MS. The size and location of this plaque likely resulted in clinical deficits in the limb/s supplied by nerve fibers in these tracts. (Material courtesy of Dr. G. Zurhein)

chronic MS plaques and therefore may be targets for drug delivery using molecules known to promote oligodendrocyte differentiation or survival in vitro. However, as of yet there has been a lack of direct evidence that such an approach is feasible, that molecules can be directed only to the CNS, or that such molecules if delivered parenterally, will not negatively affect cells in other organs. Hence, I will consider only the challenges to remyelination by glial cell transplantation, an approach that has been extensively explored in the last decade (Franklin and ffrench-Constant 1997a; Duncan 1996; Duncan et al. 1997; Blakemore et al. 1995a; Fig. 1).

7.2 Why Does Remyelination Fail in Chronic MS and How Does This Relate to Attempts at Repair?

There are many possible reasons why endogenous remyelination finally fails in patients with relapsing-remitting disease. These are summarized below.

1. Lack of mature oligodendrocytes
2. Failure of adult progenitors to respond
3. Inhibition of remyelination by cytokines OR lack of inflammation inhibits remyelination
4. Astrocyte scar acts as barrier
5. Chronically demyelinated axons lack myelinating signal
6. Acute or chronic loss of axons

While the initial immune-mediated attack may be on myelin, mature oligodendrocytes are usually lost in acute areas of demyelination (Ozawa et al. 1994). However, oligodendrocyte progenitors or cells of the early oligodendrocyte lineage may be found in chronic plaques (Wolswijk 1998; Chang et al. 2000, 2002), in which there is no evidence of remyelination. This raises the question as to whether there are inhibitory signals to repair in such areas and so implanting exogenous cells would face the same barriers. However, it should be noted that it is not possible to state that MS plaques in which oligodendrocytes or their progenitors are found would not have gone onto repair spontaneously, albeit slowly. It may also be the case that exogenous oligodendrocyte or neural stem cells transplanted into chronic plaques have greater inherent repair capabilities than host cells.

Whether inflammation is helpful or harmful to remyelination remains an important topic of ongoing research. It is clear that endogenous remyelination occurs early in disease, at a time when inflammation is present. In chronic MS, inflammation is much less obvious although remnants of an inflammatory response may always be present. An important cytokine, TNF, that has been thought to be injurious to oligodendrocyte function and survival, may actually be involved in remyelination or may play dual roles. For example, in the TNFa knock-out mouse, remyelination following cuprizone-induced demyelination is much less efficient than in wild type (Arnett et al. 2001). Microarray analysis of tissue from these mice further detailed the genes upregulated during demyelination and remyelination and showed that MHCII expression increased, raising the possibility that MHCII may be involved in repair (Arnett et al. 2003). Similar analyses have been performed in MS tissue from both acute and chronic lesions (John et al. 2002). Other experimental evidence would suggest that inflammation may promote remyelination (Blakemore et al. 1995b). From the context of transplant-induced repair, the key question is whether transplanting cells into an inflammatory milieu will result in their death, or in contrast may even be required if such an area is going to remyelinate spontaneously.

In chronic MS, plaques consist of demyelinated axons in a bed of hypertrophic astrocytic processes that are rich in glial filaments i.e., a gliotic plaque. It has been assumed that such areas of gliosis are a barrier to repair and indeed in very chronic disease, this may be so. However, astrocytes are known to produce factors that are important for myelination; hence remyelination is poor or fails in their absence (Franklin et al. 1993). Therefore, there may be a "happy medium" in which gliosis is not an impediment to repair but a time at which it indeed prevents the colonization of chronic plaques by cells that are trying to migrate toward demyelinated axons. We have shown that transplantation can result in myelination of gliotic areas of a myelin mutant CNS, though this may be less efficient than transplantation at earlier ages where gliosis is not present (Archer et al. 1997). Further research on remyelination in models in which gliosis is prominent are warranted.

A critical issue in remyelination is whether axons that have been demyelinated or nonmyelinated for months or years can actually be repaired. It is known that cues for ensheathment and myelin formation are present on the axon. What is not entirely clear however, is whether such signals persist with time. Transplant studies in mature animals and those that have had persistent areas of myelin loss or absence suggest that this may not be a problem (Archer et al. 1997). Other work suggests that CNS axons can indeed by myelinated throughout life (Setzu et al. 2004). However, the key question is whether this is the case in MS. It has been suggested that expression on axons of certain molecules that are normally developmentally regulated, such as PSA-NCAM (Charles et al. 2002) may be responsible for the failure of remyelination in MS.

Finally, the question of axon survival is critical to future attempts at promoting remyelination and restoring function (Zhang and Duncan 1999). It would be meaningless to transplant glial cells into a plaque that has only a few axons remaining. It should be noted however that plasticity of axons within the CNS may allow for persistent function or restoration to occur in the face of a major loss of axons. Therefore, attempts to remyelinate areas in which there has been a significant axonal loss may nonetheless be important to long-term functional stability and recovery. The critical issue will be the ability to quantitate, in vivo, the percentage of surviving axons in a demyelinated plaque and know whether their remyelination will lead to functional improvement.

7.3 Does Remyelination Restore Function to Axons?

A crucial question in restoration of function concerns data that supports remyelination of demyelinated axons, either endogenously or by transplanted cells, resulting in the return of normal nerve conduction. In the first instance, seminal work by Smith and colleagues has clearly shown that endogenous oligodendrocyte remyelination restores conduction in the cat CNS following focal injection of lysolecithin (Smith et al. 1981). Nerve conduction velocity and other aspects of impulse transmission returned to normal with time, despite the fact that many remyelinated axons had thin myelin sheaths and short internodes. Smith et al. have also shown that Schwann cells that frequently invade the CNS in many pathological states (Duncan and Hoffman 1997) can restore physiologic function in the rat CNS, although nerve conduction may remain somewhat slowed. These data therefore support the claim that remyelination is important in restoring function to central axons. It should also be noted however that restoration of conduction in demyelinated axons can occur without remyelination through mechanisms of sodium channel reorganization (Shrager et al. 2004). Nonetheless, as will become apparent, such a mechanism will likely not prevent axons from being targets for destruction and so in MS, may be only a temporary solution.

7.4 Does Transplantation of Myelinating Cells Restore Conduction and Have Behavioral Effects?

If spontaneous remyelination by oligodendrocytes or Schwann cells fails, then exogenous cells transplanted focally or in a disseminated fashion are an alternative to promote remyelination. In addition to promoting structural repair, this must result in return of physiologic function for this approach to be meaningful. Kocsis and colleagues in a series of experiments have provided evidence that a wide variety of myelinating cell types could improve function in experimental models in which dysmyelination or demyelination were targeted for transplant repair. In their first study, they showed that focal remyelination of the dorsal column of myelin deficient (md) rats enhanced impulse conduction with the return to near-normal levels of nerve conduction velocity (Utzschneider et al. 1994), the most important physiologic deficit in this severely dysmyelinated mutant (Utzschneider et al. 1992). This demonstration was followed by similar data in focally demyelinated lesions in adult rats where either rodent or human Schwann cells (Honmou et al. 1996; Kohama et al. 2001) or olfactory bulb ensheathing cells (Imaizumi et al. 1998) were transplanted. Interestingly, they have also shown that bone marrow-derived cells transplanted either focally (Akiyama et al. 2002) or intravenously could differentiate into cells with similarities to Schwann cells and that myelin produced by these cells also resulted in restoration of conduction. Most recently this group has modified their approach and shown, using in vivo methods, that cell transplant-induced remyelination restores conduction (Akiyama et al. 2004).

While these physiologic data are very important, it is crucial to demonstrate that remyelination resulting from transplantation restores behavioral function in the experimental model used. At present, with one exception, this has only been testable in focally induced demyelination models. Jefferey and Blakemore (1996) explored this question by injecting the gliotoxin, ethidium-bromide, into the dorsal column of the cervical cord of young rats. They showed that the demyelinating lesions resulted in motor deficits. Similar injections in the thoraco- lumbar cord do not usually result in neurologic dysfunction. They then showed that spontaneous remyelination by Schwann cells and oligodendrocytes gradually resulted in functional recovery. Inhibition of remyelination by prior X-irradiation prevented remyelination and behavioral recovery from occurring. They next explored whether transplantation of oligodendrocyte progenitors into this model would have similar results (Jeffery et al. 1999). Rats in which axonal degeneration was minimal recovered normal function. However a number of rats showed severe axonal degeneration with no recovery. Thus a positive behavioral outcome following transplantation requires confirmation in other studies.

Transplantation into the myelin mutants provides greater challenges to proving that myelination produced by exogenous cells will restore clinical function, as theoretically, much of the CNS would have to be repaired. In a single study in the shiverer (shi) mouse, transplantation of a neural stem cell line, C17-2, was demonstrated to reduce tremor in some mice (Yandava et al. 1999). This has not been repeated with cells that may be more likely to be used clinically, and one study using neurospheres as a source of oligodendrocytes transplanted into shi produced extensive myelination, but no return of function (Mitome et al. 2001).

In MS, while repair of focal lesions in certain patients may be clinically useful, the long term goal will be to repair widely scattered lesions throughout the neuraxis. Pluchino et al. (Pluchino et al. 2003) explored this possibility in an experimental allergic encephalomyelitis (EAE) model in the mouse. Affected mice were injected intravenously or intrathecally with neural stem cells derived from mouse neurospheres. In detailed analysis of these mice, they showed that treated mice had behavioral improvement. Whether this was the result of remyelination or from the transplanted cells producing growth factors or other trophic support for endogenous remyelination needs further definition.

7.5 The Extent of Myelination Achieved by Focal Transplantation

This topic has been reviewed at length recently and will only be discussed briefly here (Duncan et al. 1997; Zhang and Duncan 1999; Duncan and Milward 1995). The focal myelinating capacity of implanted cells has been tested, in general, in two model systems. In the first, cells are transplanted into the brain or spinal cord of one of the myelin mutants, animals, usually rodents that have mutations in known or unknown myelin genes. This is usually carried out in the early neonatal period as the mutants frequently have a shortened lifespan or the myelin deficiency is worst at birth. In some cases, transplants in adults have been performed, a situation more akin to clinical practice, especially in relationship to MS. The results of such transplants can be striking and can show myelination or remyelination of all axons at such focal sites, with longitudinal repair in a rostral and caudal direction (Figs. 2, 3). Such results have been found in several mutant animals and using a variety of cell types (see Sect. 7.6). In the second approach, focal areas of demyelination are created by the injection of myelinotoxic chemicals such as ethidium bromide (Blakemore et al. 1983, 1995a). Prior X-irradiation of the injection site prevents host repair and leaves a focal area of demyelination

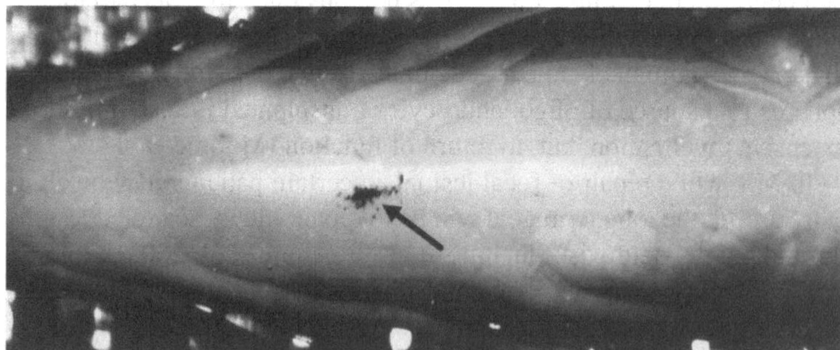

Fig. 2. View of the dorsal column of an md rat at 24 days of age that had been transplanted with oligodendrocyte progenitors 7 days earlier. The arrow marks charcoal used to identify the site of implantation of the cells. The white strip along the dorsal column is myelin made by the transplanted cells (Figure reprinted with permission of the Journal of Neuroimmunology)

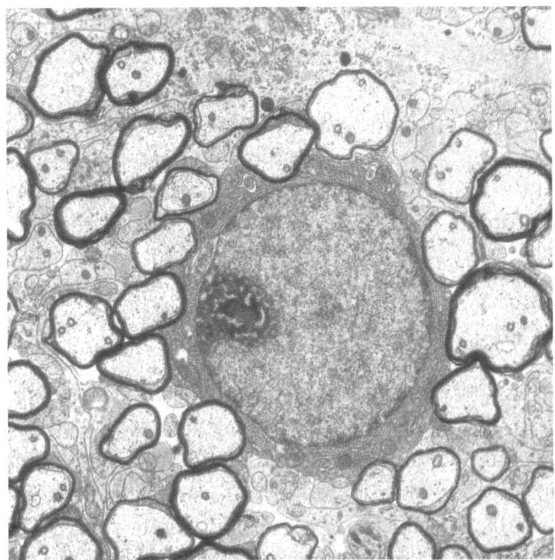

Fig. 3. An electron micrograph from an md rat that had been transplanted with the CG-4 oligodendrocyte progenitor cell line. The oligodendrocyte in the center of the field has likely myelinated the adjacent axons, most of which however, have thin myelin sheaths

into which cells can be transplanted. As in mutants receiving focal transplants of myelinating cells, these areas of demyelination can be fully repaired by implanting oligodendrocyte progenitors or other cells with myelinating capacity.

More recently, and of great relevance to MS, studies on transplantation of neural stem cells or oligodendrocyte progenitors into EAE have been performed. In the case of oligodendrocyte progenitors, it was shown that the cell line, CG4, migrated further through the neuropil in rats with EAE than in similar transplants in wild-type animals (Tourbah et al. 1997). When neural stem cells derived from neurospheres were transplanted into the lateral ventricle of rats with EAE they migrated into the brain parenchyma and preferentially target areas of inflammation where they differentiated into oligodendrocytes and astrocytes (Ben-Hur et al. 2003), and attenuated the clinical course of the disease (Einstein et al.

2003). In the more recent study of Pluchino et al. (Pluchino et al. 2003) it was clearly shown that certain aspects of the inflammatory process drive the migration of cells into the brain from the circulation or the ventricles. In wild-type mice, such migration does not occur. These data are important for MS as many large plaques are found adjacent to the ventricles of the brain, suggesting that the intraventricular deposition of cells would promote their targeting such areas of damage.

7.6 The Cell Type Used for Myelin Repair

This is one of the most critical questions which requires resolution prior to attempting trials in MS patients that will lead to functional repair. While oligodendrocytes are the endogenous myelinating cell of the CNS, other candidates, including Schwann cells and olfactory bulb ensheathing cells, should also be considered. In the case of Schwann cells, these cells are frequently seen in MS lesions and in experimental demyelinating diseases, and are commonly found to intermingle with oligodendrocytes, leading to both peripheral and central myelin repair (Duncan and Hoffman 1997). To date, the only clinical trial of cell transplantation in MS patients used autografts of peripheral nerve dissociated from the patient. (T. Volmer, personal communication) Schwann cells have remarkable capacity to myelinate central axons yet there is controversy regarding their interaction with astrocytes and this issue needs to be resolved. While it has been presumed that Schwann cells found spontaneously invading the CNS came from the adjacent nerve roots, this view has been challenged by the finding that transplanted PSA-NCAM^{+} neural stem cells could give rise to myelinating Schwann cells (Keirstead et al. 1999). Thus, the possibility exists that resident neural stem cells can differentiate in vivo into Schwann cells, given the appropriate signals. Transplantation of Schwann cells into a variety of myelin disorders has clearly shown that these cells are capable of extensive myelination, but the question of interaction with astrocytes remains unresolved (Franklin et al. 1992). Olfactory ensheathing cells offer a third alternative but will only be mentioned briefly as they have been the topic of new data and recent reviews (Franklin and Barnett 1997b; Barnett et al. 2000). While they may not be inhibited in migration and myelination

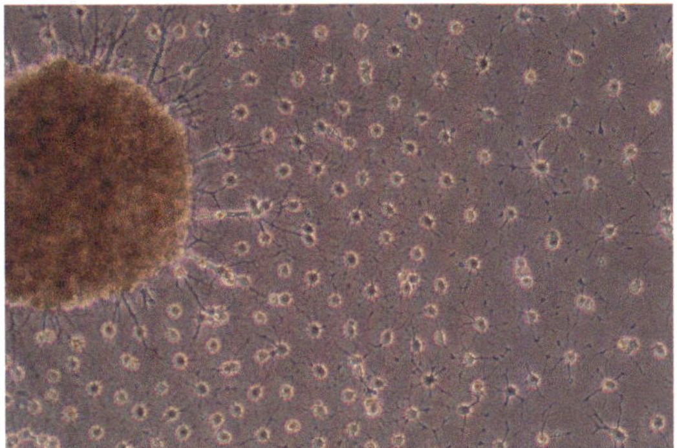

Fig. 4. An oligosphere derived from neonatal rat optic nerve growing in media containing B-104 conditioned media. Cells migrating out of the sphere appear like oligodendrocyte progenitors with a clear bipolar morphology. Cells that have moved away from the sphere have a uniform morphology of oligodendrocytes, demonstrating early stages of process outgrowth and extension

by endogenous astrocytes, there is dispute about whether they can in fact differentiate into a myelin-producing cell (Plant et al. 2002; Boyd et al. 2004).

We have focused attention on the transplantation of cells of the oligodendrocyte lineage, ranging from stem cells that can be coaxed along or through neural stem cell stages to oligodendrocyte progenitors and then on to become immature and mature oligodendrocytes. We initially showed that FACS could be used to select mature oligodendrocytes (Duncan et al. 1992) but have since that time concentrated on isolating oligodendrocyte progenitors either derived from neurospheres (Ham mang et al. 1997) or spherical collections of cells, coined oligospheres (Fig. 4) by Avellana-Adalid et al. 1996. Oligodendrocyte progenitors have a characteristic bipolar morphology and when cultured in the presence of appropriate mitogens such as PDGF a, will continue to divide and remain undifferentiated. We have used the neuroblastoma B104-conditioned media as a source of mitogens and the effect of this media on oligodendrocyte progenitors are seen in Fig. 5. These cells can be easily transduced with a retroviral vector expressing the ß-galactosidase gene, allowing them to be tracked in vivo after transplantation (Tontsch

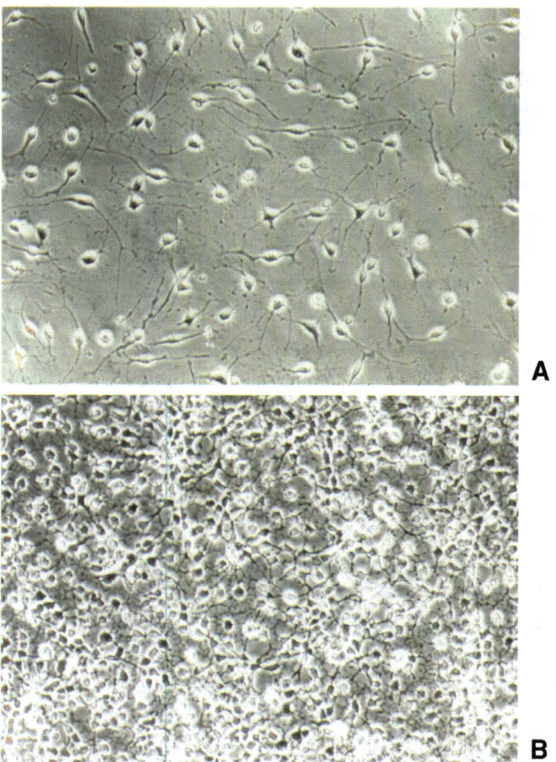

Fig. 5A, B. The proliferative capacity of oligodendrocyte progenitors in vitro when exposed to the appropriate mitogen. CG-4 cells grown in the presence of B-104 conditioned media, early in culture are phase bright and bipolar (**A**). Within a few days however, the same cover slip shows that the cells have increased in number and have become confluent (**B**). Passaging these cells allows for continued propagation of immature cells

et al. 1994). We have also labeled such cells with paramagnetic nanoparticles which allows their migration to be tracked both ex vivo (Bulte et al. 1999) and in vivo (Bulte et al. 2001).

Oligodendrocytes can also be derived from mouse embryonic stem (ES) cells in vitro and when transplanted into the md rat nondifferentiated neural stem cells give rise to myelinating oligodendrocytes (Brustle et al. 1999). As a potential source of repair in MS, human ES cells (Thomson et al. 1998) offer great potential as a infinite source of cells for treating many neurologic diseases. Unlike mouse ES cells, human ES cells do not differentiate readily into oligodendrocytes in vitro and in vivo (Zhang et al. 2001). In the presence of bFGF, human ES cells that form embryoid bodies will differentiate into neuroepithelia. From these cells, neurons and astrocytes can be cultured, but only a minor proportion become oligodendrocytes (Zhang et al. 2001). At present, therefore, the culture conditions that promote oligodendrocyte differentiation in mouse ES cells differ from human. This is not surprising as growth factors and mitogens that are biologically active on rodent cells frequently differ in effect on human cells. While ES cells may be the most promising source of cells for brain repair, oligodendrocyte progenitors derived from the fetal or adult human brain are also a candidate. Windrem et al. (Windrem et al. 2004) have shown that FACS sorting for the surface antigen A2B5 can yield pure populations of oligodendrocyte progenitors from fetal or adult human brain that can myelinate shi mouse axons on transplantation (Windrem et al. 2004). Surprisingly, however, oligodendrocyte progenitors from the adult brain myelinated more efficiently than fetal cells, but were isolated in lesser numbers, making human fetal oligodendrocyte progenitors a more likely practical source for treating human myelin disease.

7.7 Conclusions

Glial cell transplantation appears on the edge of being translated from a purely experimental approach into clinical application. Certain key questions remain however: what cell will be used, and in MS, what patient population and target sites will be selected? Answers to these questions are being eagerly sought and clinical advances should occur within the next 10 years. Repair of myelin sheaths offers the dual advantage of neuroprotection along with restoration of axonal function.

Acknowledgements. The work from my laboratory cited here has been carried out by many skilled investigators to whom I am most appreciative. This work has been supported by the NIH, Myelin Project, Hamilton Roddis Foundation,

Elizabeth Elser Doolittle Trust and Oscar Rennebohm Foundation. I am grateful to Ingrid Hertel for her help in preparing this chapter.

References

Akiyama Y, Radtke C, Kocsis JD (2002) Remyelination of the rat spinal cord by transplantation of identified bone marrow stromal cells. J Neurosci 22:6623–6630

Akiyama Y, Lankford KL, Radtke C, Greer CA, Kocsis JD (2004) Remyelination of spinal cord axons by olfactory ensheathing cells and Schwann cells derived from a transgenic rat expressing alkaline phosphatase marker gene. Neuron Glia Biol 1:1–9

Archer DR, Cuddon PA, Lipsitz D, Duncan ID (1997) Myelination of the canine central nervous system by glial cell transplantation: a model for repair of human myelin disease. Nat Med 3:54–59

Arnett HA, Mason J, Marino M, Suzuki K, Matsushima GK, Ting JP (2001) TNF? promotes proliferation of oligodendrocyte progenitors and remyelination. Nat Neurosci 4:1116–1122

Arnett HA, Wang Y, Matsushima GK, Suzuki K, Ting JPY (2003) Functional genomic analysis of remyelination reveals importance of inflammation in oligodendrocyte regeneration. J Neurosci 23:9824–9832

Avellana-Adalid V, Lachapelle F, Baron-Van Evercooren A (1996) Expansion of rat oligodendrocyte precursors into proliferate oligospheres that retain their myelinating capacity after transplantation into the shiverer newborn brain. J Neurosci Res 45:558–570

Barnett SC, Alexander CL, Iwashita Y, Gilson JM, Crowther J, Clark L, Dunn LT, Papanastassiou V, Kennedy PG, Franklin RJ (2000) Identification of a human olfactory ensheathing cell that can effect transplant-mediated remyelination of demyelinated CNS axons. Brain 123:1581–1588

Ben-Hur T, Einstein O, Mizrachi-Kol R, Ben-Menachem O, Reinhartz E, Karussis D, Abramsky O (2003) Transplanted multipotential neural precursor cells migrate into the inflamed white matter in response to experimental autoimmune encephalomyelitis. GLIA 41:73–80

Bitsch A, Schuchardt J, Bunkowski S, Kuhlmann T, Brück W (2000) Acute axonal injury in multiple sclerosis—correlation with demyelination and inflammation. Brain 123:1174–1183

Bjartmar C, Trapp BD (2001) Axonal and neuronal degeneration in multiple sclerosis: mechanisms and functional consequences. Curr Opin Neurol 14:271–278

Blakemore WF, Crang AJ, Evans RJ (1983) The effect of chemical injury on oligodendrocytes. In: Cuzner ML, Kelly RE (eds) Viruses and demyelinating diseases. Academic Press, London, pp 167–190

Blakemore WF, Crang AJ, Franklin RJM (1995a) Transplantation of glial cells. In: Ransom BR, Kettenmann H (eds) Neuroglial cells. Oxford University Press, Cambridge, pp 869–882

Blakemore WF, Crang AJ, Franklin RJM, Tang K, Ryder S (1995b) Glial cell transplants that are subsequently rejected can be used to influence regeneration of glial cell environments in the CNS. GLIA 13:79–91

Boyd JG, Lee J, Skihar V, Doucette R, Kawaja MD (2004) LacZ- expressing olfactory ensheathing cells do not associate with myelinated axons after implantation into the compressed spinal cord. Proc Natl Acad Sci 101:2162–2166

Brück W, Kuhlmann T, Stadelmann C (2003) Remyelination in multiple sclerosis. J Neurol Sci 206:181–185

Brustle O, Jones KN, Learish RD, Karram K, Choudhary K, Wiestler OD, Duncan ID, McKay RD (1999) Embryonic stem cell-derived glial precursors: A source of myelinating transplants. Science 285:754–756

Bulte JW, Zhang S, van Gelderen P, Herynek V, Jordan EK, Duncan ID, Frank JA (1999) Neurotransplantation of magnetically labeled oligodendrocyte progenitors: MR tracking of cell migration and myelination. Proc Natl Acad Sci Dec 96(26):15256-15261

Bulte JW, Douglas T, Witwer B, Zhang SC, Strable E, Lewis BK, Zywicke H, Miller B, van Gelderen P, Moskowitz BM, Duncan ID, Frank JA (2001) Magnetodendrimers allow endosomal magnetic labeling and in vivo tracking of stem cells. Nat Biotechnol 19:1141–1147

Chang A, Nishiyama A, Peterson J, Prineas J, Trapp BD (2000) NG2- positive oligodendrocyte progenitor cells in adult human brain and multiple sclerosis lesions. J Neurosci 20:6404–6412

Chang A, Tourtellotte WW, Rudick R, Trapp BD (2002) Premyelinating oligodendrocytes in chronic lesions of multiple sclerosis. N Engl J Med 346:165–173

Charles P, Reynolds R, Seilhean D, Rougon G, Aigrot MS, Niezgoda A, Zalc B, Lubetzki C (2002) Re-expression of PSA-NCAM by demyelinated axons: an inhibitor of remyelination in multiple sclerosis? Brain 125:1972–1979

Duncan ID (1996) Glial cell transplantation and remyelination of the CNS. Neuropathol Appl Neurobiol 22:87–100

Duncan ID, Milward EA (1995) Glial cell transplants: Experimental therapies of myelin diseases. Brain Pathol 5:301–310

Duncan ID, Hoffman RL (1997) Schwann cell invasion of the central nervous system of the myelin mutants. J Anat 190:35–49

Duncan ID, Paino C, Archer DR, Wood PM (1992) Functional capacities of transplanted cell-sorted adult oligodendrocytes. Dev Neurosci 14:114–122

Duncan ID, Grever WE, Zhang S-C (1997) Repair of myelin disease: strategies and progress in animal models. Mol Med Today 3:554–561

Einstein O, Karussis D, Grigoriadis N, Mizrachi-Kol R, Reinhartz E, Abramsky O, Ben-Hur T (2003) Intraventricular transplantation of neural precursor cell spheres attenuates acute experimental allergic encephalomyelitis. Mol Cell Neurosci 27:1074–1082

Ferguson B, Matyszak MK, Esiri MM, Perry VH (1997) Axonal damage in acute multiple sclerosis lesions. Brain 120:393–399

Franklin RJM, ffrench-Constant C (1997a) Molecular biology of multiple sclerosis. In: Russell WC (ed) Transplantation and repair in multiple sclerosis. Wiley, New York, pp 231–242

Franklin RJM, Barnett SC (1997b) Do olfactory glia have advantages over Schwann cells for CNS repair. J Neurosci Res 50:665–672

Franklin RJM, Crang AJ, Blakemore WF (1992) Type 1 astrocytes fail to inhibit Schwann cell remyelination of CNS axons in the absence of cells of the O-2A lineage. Dev Neurosci 14:85–92

Franklin RJM, Crang AJ, Blakemore WF (1993) The role of astrocytes in the remyelination of glia-free areas of demyelination. In: Seil FJ (ed) Advances in neurology. Raven Press, New York, pp. 125–133

Hammang JP, Archer DR, Duncan ID (1997) Myelination following transplantation of EGF-responsive neural stem cells into a myelin-deficient environment. Exp Neurol 147:84–95

Honmou O, Kocsis JD, Waxman SG, Felts PA (1996) Restoration of normal conduction properties in demyelinated spinal cord axons in the adult rat by transplantation of exogenous Schwann cells. J Neurosci 16:3199–3208

Imaizumi T, Lankford KL, Waxman SG, Greer CA, Kocsis JD (1998) Transplanted olfactory ensheathing cells remyelinate and enhance axonal conduction in the demyelinated dorsal columns of the rat spinal cord. J Neurosci 18:6176–6185

Jeffery ND, Blakemore WF (1996) Locomotor deficits induced by experimental spinal cord demyelination are abolished by spontaneous remyelination. Brain 119:101–111

Jeffery ND, Crang AJ, O'Leary MT, Hodge SJ, Blakemore WF (1999) Behavioural consequences of oligodendrocyte progenitor cell transplantation into experimental demyelinating lesions in the rat spinal cord. Eur J Neurosci 11:1508–1514

John GR, Shankar SL, Shafit-Zagardo B, Massimi A, Lee SC, Raine CS, Brosnan CF (2002) Multiple sclerosis: Re-expression of a developmental pathway that restricts oligodendrocyte maturation. Nat Med 8:1115–1121

Keirstead HS, Ben-Hur T, Rogister B, O'Leary MT, Dubois-Dalcq M, Blakemore WF (1999) Polysialylated neural cell adhesion molecule-positive CNS precursors generate both oligodendrocytes and Schwann cells to remyelinate the CNS after transplantation. J Neurosci 19:7529–7536

Kohama I, Lankford KL, Preiningerova J, White FA, Vollmer TL, Kocsis JD (2001) Transplantation of cryopreserved adult human Schwann cells enhances axonal conduction in demyelinated spinal cord. J Neurosci 21:944–950

Kornek B, Lassmann H (1999) Axonal pathology in multiple sclerosis. A historical note. Brain Pathol 9:651–656

Lassmann H, Brück W, Lucchinetti C, Rodriguez M (1997) Remyelination in multiple sclerosis. Multiple Sclerosis 3:133–136

Mitome M, Low HP, van den Pol A, Nunnari JJ, Wolf MK, Billings- Gagliardi S, Schwartz WJ (2001) Towards the reconstruction of central nervous system white matter using neural precursor cells. Brain 124:2147–2161

Ozawa K, Suchanek G, Breitschopf H, Bruck W, Budka H, Jellinger K, Lassmann H (1994) Patterns of oligodendroglia pathology in multiple sclerosis. Brain 117:1311–1322

Plant GW, Currier PF, Cuervo EP, Bates ML, Pressman Y, Bunge MB, Wood PM (2002) Purified adult ensheathing glia fail to myelinate axons under culture conditions that enable Schwann cells to form myelin. J Neurosci 22:6083–6091

Pluchino S, Quattrini A, Brambilla E, Gritti A, Salani G, Dina G, Galli R, Del Carro U, Amadio S, Bergami A, Furlan R, Comi G, Vescovi AL, Martino G (2003) Injection of adult neurospheres induces recovery in a chronic model of multiple sclerosis. Nature 422:688–694

Setzu A, ffrench-Constant C, Franklin RJM (2004) CNS axons retain their competence for myelination throughout life. GLIA 45:307–311

Shrager P, Simon W, Kazarinova-Noyes K (2004) Remyelination and restoration of axonal function by glial cell transplantation. In: Waxman SG (ed) Multiple sclerosis as a neuronal disease. Academic Press, San Diego

Smith KJ, Blakemore WF, McDonald WI (1981) The restoration of conduction by central remyelination. Brain 104:383–404

Thomson JA, Itskovitz-Eldor J, Shapiro SS, Waknitz MA, Swiergiel JJ, Marshall VS, Jones JM (1998) Embryonic stem cell lines derived from human blastocysts. Science 282:1145-1147

Tontsch U, Archer DR, Dubois-Dalcq M, Duncan ID (1994) Transplantation of an oligodendrocyte cell line leading to extensive myelination. Proc Natl Acad Sci 91:11616–11620

Tourbah A, Linnington C, Bachelin C, Avellana-Adalid V, Wekerle H, Baron-Van Evercooren A (1997) Inflammation promotes survival and migration of the CG4 oligodendrocyte progenitors transplanted in the spinal cord of both inflammatory and demyelinated EAE rats. J Neurosci Res 50:853–861

Trapp BD, Peterson J, Ransohoff RM, Rudick R, Mork S, Bo L (1998) Axonal transection in the lesions of multiple sclerosis. N Engl J Med 338:278–285
Utzschneider D, Black JA, Kocsis JD (1992) Conduction properties of spinal cord axons in the myelin- deficient rat mutant. Neuroscience 49:221– 228
Utzschneider DA, Archer DR, Kocsis JD, Waxman SG, Duncan ID (1994) Transplantation of glial cells enhances action potential conduction of amyelinated spinal cord axons in the myelin-deficient rat. Proc Natl Acad Sci 91:53–57
Windrem MS, Nunes MC, Rashbaum WK, Schwartz TH, Goodman RA, McKhann G 2nd, Roy NS, Goldman SA (2004) Fetal and adult human oligodendrocyte progenitor cell isolates myelinate the congenitally dysmyelinated brain. Nat Med 10(1):93–97
Wolswijk G (1998) Chronic stage multiple sclerosis lesions contain a relatively quiescent population of oligodendrocyte precursor cells. J Neurosci 18:601–609
Yandava BD, Billinghurst LL, Snyder EY (1999) "Global" cell replacement is feasible via neural stem cell transplantation: Evidence from the dysmyelinated shiverer mouse brain. Proc Natl Acad Sci 96:7029–7034
Zhang S-C, Duncan ID (1999) Functional neural transplantation. Dunnett SB, Bjorklund A (eds) Elsevier, Amsterdam
Zhang S-C, Lundberg C, Lipsitz D, O'Connor LT, Duncan ID (1998a) Generation of oligodendroglial progenitors from neural stem cells. J Neurocytol 27:475–489
Zhang S-C, Lipsitz D, Duncan ID (1998b) Self-renewing canine oligodendroglial progenitor expanded as oligospheres. J Neurosci Res 54:181– 190
Zhang S-C, Wernig M, Duncan ID, Brüstle O, Thomson JA (2001) In vitro differentiation of transplantable neural precursors from human embryonic stem cells. Nat Biotechnol 19:1129–1133

8 Gene and Stem Cell Therapy for Autoimmune Demyelination

S. Pluchino, M. Bacigaluppi, S. Bucello, E. Butti, M. Deleidi, L. Zanotti, G. Martino, R. Furlan

8.1 Introduction

The pathological hallmark of multiple sclerosis (MS) is the presence within the central nervous system (CNS) of inflammatory infiltrates containing few autoreactive T cells and a multitude of pathogenic nonspecific lymphocytes determining patchy demyelination, axonal loss, and severe glial scarring. It is currently believed that CNS antigen-specific T cells provide the organ specificity of the pathogenic process and regulate the influx within the CNS of nonantigen- specific mononuclear cells that, in turn, act as effector cells by directly destroying oligodendrocytes and/or by releasing myelinotoxic substances (Kieseier et al. 1999; Martino and Hartung 1999). In most instances, however, oligodendrocytes or their precursors are morphologically preserved in demyelinating plaques during the early phase of the disease thus remaining capable of differentiation and remyelination (Lucchinetti et al. 1996; Martino and Hartung 1999; Prineas et al. 1993). A successful therapeutic approach of MS should therefore be aimed to (a) inhibit the activation of antigen- and nonantigen-specific immune cells, and/or (b) to rescue the surviving oligodendrocytes within demyelinating plaques, or both. CNS gene delivery using nonreplicative viral vectors able to infect postmitotic cells such as those resident in the CNS has being proposed as a useful therapeutic approach to deliver anti-inflammatory cytokine and – possibly – growth factor genes for either inhibiting the activation of antigen- and nonantigen-specific immune cells (anti-inflammatory therapies) or fostering surviving oligodendrocyte progenitors to differentiate into myelin forming cells ("remyelinating" therapies), respectively (Furlan et al. 2003). Experimental cell- based strategies aimed at replacing damaged myelin-forming cells have also been developed in the last few years (Pluchino et al. 2004). Most of these therapeutic approaches have appeared as unfeasible, owing to the mutifocality of the demyelinating process in MS patients and the inability of in vitro growth and of differentiation of large numbers of myelin-forming cells. However, recently proposed adult somatic stem cell-based therapies (Pluchino et al. 2003) have partially overcome these limitations.

Recent advances in the development of gene- and stem cell-based therapies, aimed at promoting multifocal remyelination, are here discussed.

8.2 CNS Delivery of Neuroprotective Genes

8.2.1 Anti-inflammatory Cytokine Genes

Proinflammatory cytokines (i.e., Th1 cytokines such as interferon (IFN)γ, tumor necrosis factor (TNF)α/β, IL-2) are believed to play a crucial role in MS pathogenic process since they can promote and sustain the development of myelin-specific T cells and the recruitment, within the CNS, of peripheral myelinotoxic effector cells, i.e., monocyte/macrophages (Kieseier et al. 1999). In addition, proinflammatory cytokines such as TNFα can be directly toxic for oligodendrocytes (Selmaj et al. 1998). Proinflammatory cytokines thus represent a suitable therapeutic target in MS. However, systemic delivery of anti- inflammatory cytokines has shown limited or no efficacy, and considerable toxicity when administered to MS patients. Intramuscular or subcutaneous administration of interferon IFNβ reduces by only one third the exacerbation rate of the disease, does not change substantially the progression of disability, and induces the production of neutralizing anti-IFNβ antibodies (The INFB Multiple Sclerosis Study Group 1993; 1996). Systemic administration of TGFβ fails to prevent the appearance of new inflammatory magnetic resonance imaging lesions in the brain of patients and causes reversible nephrotoxicity (Calabresi et al. 1998). Intravenous injection of the recombinant TNF-receptor p55 immunoglobulin fusion protein Lenercept® increases the number of patients experiencing disease relapses and induces the formation of neutralizing antibodies (The Lenercept Multiple Sclerosis Study Group 1999). These disappointing results are possibly due to the scarce capacity of systemically administered cytokines to cross the blood brain barrier (BBB) and accumulate in the CNS where the MS pathogenic process takes place (Khan et al. 1996). Moreover, cytokines act in an autocrine-paracrine fashion and have a short half-life thus being consumed at the site of production and/or administration. The CNS delivery of anti-inflammatory cytokines (i.e., Th2 cytokines such as IL-4, IL-5, IL-10, IL-13) could partially overcome these limitations. Biological vectors engineered with heterologous genes coding for anti-inflammatory cytokines and injected into the CNS might represent appropriate tools to deliver anti-inflammatory "protective" Th2 cytokines in inflammatory demyelinating diseases of the CNS.

8.2.2 The Ependymal Route

We have developed an alternative approach to CNS gene therapy by using the ependymal route (Martino et al. 2001). Injection into the cerebrospinal fluid (CSF) allows viral vectors to infect only cells lining liquoral spaces, like ependymal and leptomeningeal cells. The large number of viral particles that can be delivered in this relatively small compartment ensures high infection efficiency. If the delivered gene codes for a soluble, secreted molecule, this will be released into the CSF and be able to travel through the ventricular system to reach all brain areas, remaining confined to the brain and unable to induce unwanted side effects in the periphery. This approach has been used in EAE in mice and nonhuman primates (Furlan et al. 2003), and in ischemic stroke (Shimamura et al. 2004), but holds promise for all multifocal brain diseases. Since ependymal and leptomeningeal cells are slow dividing cells that are poorly renewed, their infection allows long-term expression, up to 6 months, of the delivered transgene. Summarizing, the advantages of the ependymal route for CNS gene therapy are:

1. The high concentration of soluble therapeutic proteins that can be achieved in the CSF
2. The possibility of reaching multiple brain areas and thus be useful in multifocal CNS disorders
3. The ability of vectors injected in the CSF to express the transgene long-term, being potentially useful for chronic diseases
4. The absence of unwanted peripheral toxicity and side effects
5. The absence of an intrathecal immune response towards the viral vector which should allow repetitive injections without loss of therapeutic efficacy

Some of these features rely on the nature of the protein encoded by the transferred gene. The therapeutic molecule has to be soluble and secreted, and its ability to travel across the brain–CSF barrier and "soak" the brain parenchyma depends on its physical and chemical properties, and has to be assessed for each molecule. Using an HSV-1-derived vector expressing the cytokine IFNγ in mice, we have been able to show that the biological effect (i.e., induction of MHC-I and -II expression) of the molecule transferred by gene therapy could be detected in the brain parenchyma at least at 1 mm from liquoral spaces (Furlan et al. 2001). With the same type of vector, but delivering the cytokine IL-4, however, we were able to interfere with an inflammatory disease ongoing in the brain of larger mammals such as Rhesus monkeys (Poliani et al. 2001), indicating that data obtained in mice were, most likely, underestimated. HSV-1-derived vectors were used to deliver IL-4 in mice with EAE, either before or after disease onset. I.c. injection of an IL-4-coding HSV-1 vector was able to inhibit chronic-remitting EAE development in Biozzi AB/H mice immunized with the myelin oligodendrocyte glycoprotein (MOG)40–55 peptide (Furlan et al. 1998). Disease prevention was associated to a decreased recruitment within the CNS of monocyte/macrophages from the peripheral circulation. The i.c., HSV-1- mediated, IL-4 delivery was able also to ameliorate ongoing relapsing-remitting EAE in spinal cord homogenate-immunized Biozzi AB/H mice, determining, in this case, a significant modulation of the local cytokine milieu, leading to downregulation of proinflammatory cytokines and chemokines (Furlan et al. 2001). This latter approach has been tested also in nonhuman primates affected by a very acute, invariably fatal, form of EAE induced by immunization with whole myelin. Sixty percent of monkeys, i.c.-injected at the time of disease onset with an HSV-1 vector engineered with the human IL-4 gene, were completely protected from EAE signs and symptoms (Poliani et al. 2001). The ependymal route using HSV-1-derived vectors has been employed also to deliver the IFNγ gene, which was able to both inhibit or treat MOG35–55-induced chronic EAE in C57BL/6 mice, through the induction of in situ apoptotic death of encephalitogenic T cells (Furlan et al. 2001). Using the same protocol, we used HSV-1 derived vectors to deliver the IL-1 receptor antagonist (IL-1ra) to C57BL/6 EAE mice immunized with MOG35–55 and obtained delay of disease onset and decreased severity only on a prophylactic schedule (R. Furlan, unpublished). HSV-1-derived vectors, however, have a low chance to be employed in a human clinical setting due to: (a) the short term transgene production i.e., up to 4 weeks); (b) their possible immunogenicity; and (c) their derivation from a virus potentially very dangerous for its selective neurovirulence.

8.2.3 Neurotrophic Growth Factor Genes

The development of therapies aimed to promote remyelination is a major issue in MS, where repeated episodes of demyelination over time lead to axonal loss and permanent neurological impairment (Lucchinetti et al. 1996; Prineas et al. 1993). Therapies aimed to promote myelin restoration are so far mainly based on transplantation of oligodendrocytes (or oligodendrocyte precursors) or multipotential neural stem cells (NSCs; Franklin 2002), and on the use of neurotrophic growth factors able to promote migration, proliferation, and differentiation of oligodendrocyte precursors (Franklin et al. 2001).

HSV-1-mediated intracisternal delivery of the fibroblast growth factor (FGF)-II gene was able to induce oligodendrocyte precursors proliferation and migration, thus ameliorating ongoing chronic EAE in MOG35–55-immunized C57BL/6 mice (Ruffini et al. 2001). Growth factors have been partially successful in the therapy of EAE (Chen et al. 1998; Croxford et al. 1997; Villoslada et al. 2000), but not in MS patients where their systemic administration caused heavy side effects (Calabresi et al. 1998). However, results in humans are still preliminary and questionable (i.e., natural inactive TGFβ1 was used in EAE while active TGFβ1 was used in MS patients) since there is clear in vitro evidence that growth factors may stimulate the proliferation of glial progenitor cells, and their differentiation into myelinating oligodendrocytes.

8.3 Stem Cell Therapies

Somatic stem cells represent an alternative source of cells that can be used to promote myelin repair (Pluchino et al. 2004). Encouraging preliminary results have been obtained by transplanting adult neural stem cells (aNSCs) (Fig. 1) into rodents affected by CNS demyelination (Ben-Hur et al. 2003b; Pluchino et al. 2003). However, although somatic stem cells may integrate within the CNS and possibly repair the myelin damage, further studies are required to assess the in vivo plasticity of these cells and the safety and efficacy of this therapeutic approach. Furthermore, there are some additional questions we need to confront with before prospecting any potential human application of such therapies: (a) the ideal stem cell source for transplantation; (b) the route of cell administration; (c) the differentiation and persistence of the transplanted cells into the targeted tissue; and, last but not least, (d) functional and long-lasting integration of transplanted cells into the host tissue has to be achieved.

Whatever the organ or tissue necessities, the "gold standard" cell for replacement therapies has to be inherently plastic. Stem cells can meet this criterion since they are intrinsically able either to adapt their terminal cell fate to different environmental needs (differentiation plasticity) or to transdifferentiate (developmental plasticity). Moreover, stem cells represent a potentially unlimited source of myelin-forming cells while

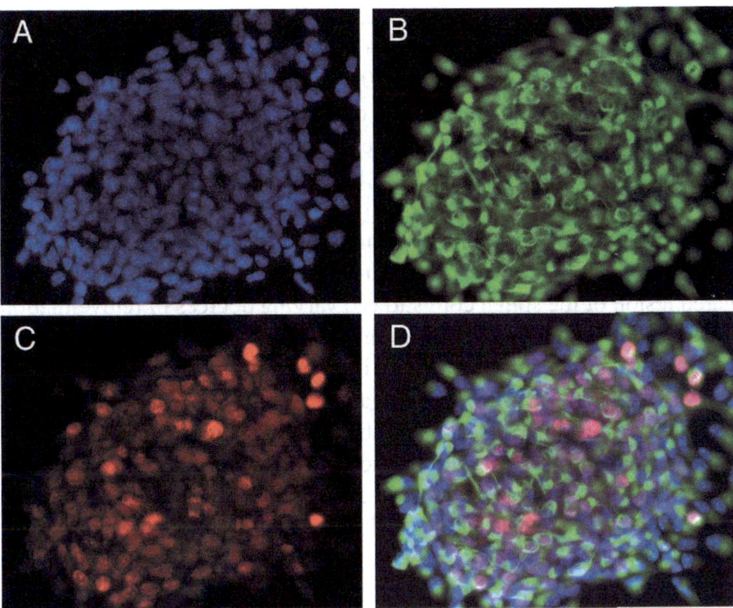

Fig. 1A–D. Adult neural stem cells (aNSC) for transplantation in CSN demyelination. Double immunofluorescence of mouse subventricular zone (SVZ)-derived aNSCs grown as neurospheres in vitro and labeled with antibodies against nestin (**B**, *green*), and the proliferation marker Ki67 (**C**, *red*). Nuclei have been counterstained with dapi (**A**, *blue*). Merged image of **A–C** is shown in **D**. ×40

either more mature or even postmitotic myelin-forming cells are difficult to manipulate and can be expanded in vitro only scarcely (Franklin 2002). Both embryonic stem cells (ES) and aNSCs might represent the ideal cell source for cell replacement-based therapies in myelin CNS disorders. aNSCs showed the potential to repair demyelinating lesions by acquiring a preferential glial cell-fate once transplanted in vivo into rodents suffering from either acute or chronic autoimmune inflammatory demyelination (Ben-Hur et al. 2003b; Bulte et al. 2003; Einstein et al. 2003; Pluchino et al. 2003). ES cells have been able to differentiate into glial cells and re-ensheath in vivo demyelinated axons when transplanted in animal models of either genetically determined or chemically-induced demyelination (Brustle et al. 1999; Chu et al. 2003; Liu et al. 2000; McDonald et al. 1999;Reubinoff et al. 2001; Zhang et al. 2001). However, transplantation of ES cells has been complicated by the formation of heterologous tissues and teratomas within the organ of transplantation (Brustle et al. 1997, 1999; Deacon et al. 1998; Yanai et al. 1995).

Other somatic stem cells of nonneuronal origin have been recently used to repair the myelin sheath in vivo. Rats with an acute demyelinated lesion of the spinal cord showed varying degrees of remyelination – which was proportional to the number of injected cells – after systemic or intralesional injection of acutely isolated mononuclear bone marrow-derived stem cells (BMSCs); (Akiyama et al. 2002a; Inoue et al. 2003). Moreover, bone marrow-derived stromal cells induced remyelination and improvement of axonal conduction velocity once transplanted by direct microinjection into the demyelinated spinal cord of immunosuppressed rat (Akiyama et al. 2002b). These results, although encouraging, are still too preliminary to draw any meaningful conclusion about the therapeutic use of BMSCs in demyelinating disorders.

The route of cell administration represents another key issue for stem cell transplantation in multifocal CNS diseases. While direct intralesional cell transplantation can be instrumental in focal CNS disorders (e.g., Parkinson's disease or spinal cord injury), alternative approaches have to be established in multifocal CNS disorders (e.g., MS), where multiple CNS injections would be impractical. Interestingly enough, some recent experiments have shown that stem cells e.g., bone marrow cells, mesenchymal cells, (aNSCs) may reach multiple areas of the CNS once injected into the blood stream (i.v.) or into the cerebrospinal fluid circulation (i.c.) of rodents with multifocal demyelinating disorders of inflammatory origin (Ben-Hur et al. 2003b; Einstein et al. 2003; Pluchino et al. 2003). This specific homing has been explained, at least in part, by the constitutive expression by transplanted stem cells of a wide array of inflammatory molecules such as adhesion molecules (i.e., integrins, selectins, immunoglobulins, etc.), cytokines, chemokines, and chemokine receptors (Ben- Hur et al. 2003a; Coulombel et al. 1997; Klassen et al. 2003; Luo et al. 2002; Papayannopoulou 2003; Pluchino et al. 2003; Schmid and Anton 2003).

Ideally, once in the target organ, transplanted stem cells should differentiate into the appropriate daughter cells and persist as long as needed at the site of engraftment. Very little is known about the mechanisms instructing the terminal differentiation of stem cells in vivo; however, there is strong evidence that the local micro-environment might dictate the fate choice of transplanted uncommitted stem cells. In this respect, undifferentiated multipotent aNSCs or even totipotent ES cells, transplanted in different experimental neurological conditions, have shown considerable capacity to restrict their terminal fate to tissue-specific cues and replace nonfunctioning neural cells of different lineages, including myelin-forming cells (Brustle et al. 1999; Liu et al. 2000; McDonald et al. 1999; Pluchino et al. 2003). It has been shown that also BMSCs may give rise to myelin-forming cells once transplanted in vivo into demyelinated areas (Akiyama et al. 2002a; Inoue et al. 2003). However, developmental transdifferentiation of BMSCs into neural stem cells – although clearly described – has been recently disputed by studies showing that this is a rare event in vivo and that most of "transdifferentiated" BMSCs are transplanted cells whose nuclei are fused with those of endogenous resident neural cells (Alvarez-Dolado et al. 2003; Mezey et al. 2003; Priller et al. 2001; Weimann et al. 2003).

Finally, the functional integration of stem cells at the site of homing/transplantation is the most critical issue. Although indications that stem cells – whatever their tissue of origin – can reach the target organ and differentiate into the appropriate lineage exist, there is still scarce evidence that these cells can reconstruct the 3D brain architecture and give raise to properly functioning cells integrating into the brain circuitries. So far, most studies on aNSCs or BMSCs have relied strictly on morphological or immunohistochemical evidence (Doetsch 2003). Further studies fulfilling several strict criteria are therefore necessary to determine whether a stem cell has generated a functional neuronal or glial cell.

In conclusion, the intrinsic complex nature of MS – in particular its chronicity and multifocality and the presence of both inflammatory (acute myelin and axon destruction) and degenerative (chronic axonal loss) features – poses great limitations for cell-based remyelinating therapies. Although promising results have been obtained using stem cell-based therapies in preclinical settings, the great challenge for the future is to understand how to use these cells in a reliable, safe, and reproducible fashion in order to hopefully achieve a complete functional and anatomical rescuing of myelin architecture.

References

Akiyama Y, Radtke C, Honmou O, Kocsis JD (2002a) Remyelination of the spinal cord following intravenous delivery of bone marrow cells. Glia 39:229–236

Akiyama Y, Radtke C, Kocsis JD (2002b) Remyelination of the rat spinal cord by transplantation of identified bone marrow stromal cells. J Neurosci 22:6623–6630

Alvarez-Dolado M, Pardal R, Garcia-Verdugo JM, Fike JR, Lee HO, Pfeffer K, et al. (2003) Fusion of bone-marrow-derived cells with Purkinje neurons, cardiomyocytes and hepatocytes. Nature 425:968–973

Ben-Hur T, Ben-Menachem O, Furer V, Einstein O, Mizrachi-Kol R, Grigoriadis N (2003a) Effects of proinflammatory cytokines on the growth, fate, and motility of multipotential neural precursor cells. Mol Cell Neurosci 24:623– 631

Ben-Hur T, Einstein O, Mizrachi-Kol R, Ben-Menachem O, Reinhartz E, Karussis D, et al. (2003b) Transplanted multipotential neural precursor cells migrate into the inflamed white matter in response to experimental autoimmune encephalomyelitis. Glia 41:73–80

Brustle O, Jones KN, Learish RD, Karram K, Choudhary K, Wiestler OD, et al. (1999) Embryonic stem cell-derived glial precursors: a source of myelinating transplants. Science 285:754–756

Brustle O, Spiro AC, Karram K, Choudhary K, Okabe S, McKay RD (1997) In vitro-generated neural precursors participate in mammalian brain development. Proc Natl Acad Sci 94:14809–14814

Bulte JW, Ben-Hur T, Miller BR, Mizrachi-Kol R, Einstein O, Reinhartz E, et al. (2003) MR microscopy of magnetically labeled neurospheres transplanted into the Lewis EAE rat brain. Magn Reson Med 50:201–205

Calabresi PA, Fields NS, Maloni HW, Hanham A, Carlino J, Moore J, et al. (1998) Phase 1 trial of transforming growth factor beta 2 in chronic progressive MS. Neurology 51:289–292

Chen LZ, Hochwald GM, Huang C, Dakin G, Tao H, Cheng C, et al. (1998) Gene therapy in allergic encephalomyelitis using myelin basic protein- specific T cells engineered to express latent transforming growth factor-beta1. Proc Natl Acad Sci 95:12516–12521

Chu K, Kim M, Jeong SW, Kim SU, Yoon BW (2003) Human neural stem cells can migrate, differentiate, and integrate after intravenous transplantation in adult rats with transient forebrain ischemia. Neurosci Lett 343:129–133

Coulombel L, Auffray I, Gaugler MH, Rosemblatt M (1997) Expression and function of integrins on hematopoietic progenitor cells. Acta Haematol 97:13–21

Croxford JL, O'Neill JK, Baker D (1997) Polygenic control of experimental allergic encephalomyelitis in Biozzi ABH and BALB/c mice. J Neuroimmunol 74:205–211

Deacon T, Dinsmore J, Costantini LC, Ratliff J, Isacson O (1998) Blastula-stage stem cells can differentiate into dopaminergic and serotonergic neurons after transplantation. Exp Neurol 149:28–41

Doetsch F (2003) The glial identity of neural stem cells. Nat Neurosci 6:1127–1134

Einstein O, Karussis D, Grigoriadis N, Mizrachi-Kol R, Reinhartz E, Abramsky O, et al. (2003) Intraventricular transplantation of neural precursor cell spheres attenuates acute experimental allergic encephalomyelitis. Mol Cell Neurosci 24:1074–1082

Franklin RJ (2002) Remyelination of the demyelinated CNS: the case for and against transplantation of central, peripheral and olfactory glia. Brain Res Bull 57:827–832

Franklin RJ, Hinks GL, Woodruff RH, O'Leary MT (2001) What roles do growth factors play in CNS remyelination? Prog Brain Res 132:185–193

Furlan R, Brambilla E, Ruffini F, Poliani PL, Bergami A, Marconi PC, et al. (2001) Intrathecal delivery of IFN-gamma protects C57BL/6 mice from chronic-progressive experimental autoimmune encephalomyelitis by increasing apoptosis of central nervous system-infiltrating lymphocytes. J Immunol 167:1821–1829

Furlan R, Pluchino S, Marconi PC, Martino G (2003) Cytokine gene delivery into the central nervous system using intrathecally injected nonreplicative viral vectors. Methods Mol Biol 215:279–289

Furlan R, Poliani PL, Galbiati F, Bergami A, Grimaldi LM, Comi G, et al. (1998) Central nervous system delivery of interleukin 4 by a nonreplicative herpes simplex type 1 viral vector ameliorates autoimmune demyelination. Hum Gene Ther 9:2605–2617

The IFNB Multiple Sclerosis Study Group (1993) Interferon beta-1b is effective in relapsing-remitting multiple sclerosis. I. Clinical results of a multicenter, randomized, double-blind, placebo-controlled trial. Neurology 43:655–661

The IFNB Multiple Sclerosis Study Group and the University of British Columbia MS/MRI Analysis Group (1996) Neutralizing antibodies during treatment of multiple sclerosis with interferon beta 1b: experience during the first three years. Neurology 47:889–894

Inoue M, Honmou O, Oka S, Houkin K, Hashi K, Kocsis JD (2003) Comparative analysis of remyelinating potential of focal and intravenous administration of autologous bone marrow cells into the rat demyelinated spinal cord. Glia 44:111–118

Khan OA, Xia Q, Bever CT, Jr., Johnson KP, Panitch HS, Dhib- Jalbut SS (1996) Interferon beta-1b serum levels in multiple sclerosis patients following subcutaneous administration. Neurology 46:1639–1643

Kieseier BC, Storch MK, Archelos JJ, Martino G, Hartung HP (1999) Effector pathways in immune mediated central nervous system demyelination. Curr Opin Neurol 12:323–336

Klassen HJ, Imfeld KL, Kirov, II, Tai L, Gage FH, Young MJ, et al. (2003) Expression of cytokines by multipotent neural progenitor cells. Cytokine 22:101–106

The Lenercept Multiple Sclerosis Study Group and The University of British Columbia MS/MRI Analysis Group 1999) TNF neutralization in MS: results of a randomized, placebo-controlled multicenter study. Neurology 53:457–465

Liu S, Qu Y, Stewart TJ, Howard MJ, Chakrabortty S, Holekamp TF, et al. (2000) Embryonic stem cells differentiate into oligodendrocytes and myelinate in culture and after spinal cord transplantation. Proc Natl Acad Sci 97:6126–6131

Lucchinetti CF, Bruck W, Rodriguez M, Lassmann H (1996) Distinct patterns of multiple sclerosis pathology indicates heterogeneity on pathogenesis. Brain Pathol 6:259–274

Luo Y, Cai J, Liu Y, Xue H, Chrest FJ, Wersto RP, et al. (2002) Microarray analysis of selected genes in neural stem and progenitor cells. J Neurochem 83:1481–1497

Martino G, Furlan R, Comi G, Adorini L (2001) The ependymal route to the CNS: an emerging gene-therapy approach for MS. Trends Immunol 22:483--490

Martino G, Hartung HP (1999) Immunopathogenesis of multiple sclerosis: the role of T cells. Curr Opin Neurol 12:309–321

McDonald JW, Liu XZ, Qu Y, Liu S, Mickey SK, Turetsky D, et al. (1999) Transplanted embryonic stem cells survive, differentiate and promote recovery in injured rat spinal cord. Nat Med 5:1410–1412

Mezey E, Key S, Vogelsang G, Szalayova I, Lange GD, Crain B (2003) Transplanted bone marrow generates new neurons in human brains. Proc Natl Acad Sci 100:1364–1369

Papayannopoulou T (2003) Bone marrow homing: the players, the playfield, and their evolving roles. Curr Opin Hematol 10:214–219

Pluchino S, Furlan R, Martino G (2004) Cell-based remyelinating therapies in multiple sclerosis: evidence from experimental studies. Curr Opin Neurol 17:247–255

Pluchino S, Quattrini A, Brambilla E, Gritti A, Salani G, Dina G, et al. (2003) Injection of adult neurospheres induces recovery in a chronic model of multiple sclerosis. Nature 422:688–694

Poliani PL, Brok H, Furlan R, Ruffini F, Bergami A, Desina G, et al. (2001) Delivery to the central nervous system of a nonreplicative herpes simplex type 1 vector engineered with the interleukin 4 gene protects rhesus monkeys from hyperacute autoimmune encephalomyelitis. Hum Gene Ther 12:905–920

Priller J, Persons DA, Klett FF, Kempermann G, Kreutzberg GW, Dirnagl U (2001) Neogenesis of cerebellar Purkinje neurons from gene-marked bone marrow cells in vivo. J Cell Biol 155:733–738

Prineas JW, Barnard RO, Kwon EE, Sharer LR, Cho ES (1993) Multiple sclerosis: remyelination of nascent lesions. Ann Neurol 33:137–151

Reubinoff BE, Itsykson P, Turetsky T, Pera MF, Reinhartz E, Itzik A, et al. (2001) Neural progenitors from human embryonic stem cells. Nat Biotechnol 19:1134–1140

Ruffini F, Furlan R, Poliani PL, Brambilla E, Marconi PC, Bergami A, et al. (2001) Fibroblast growth factor-II gene therapy reverts the clinical course and the pathological signs of chronic experimental autoimmune encephalomyelitis in C57BL/6 mice. Gene Ther 8:1207–1213

Schmid RS, Anton ES (2003) Role of integrins in the development of the cerebral cortex. Cereb Cortex 13:219–224

Selmaj K, Walczak A, Mycko M, Berkowicz T, Kohno T, Raine CS (1998) Suppression of experimental autoimmune encephalomyelitis with a TNF binding protein TNFbp) correlates with down-regulation of VCAM-1/VLA-4. Eur J Immunol 28:2035–2044

Shimamura M, Sato N, Oshima K, Aoki M, Kurinami H, Waguri S, et al. (2004) Novel therapeutic strategy to treat brain ischemia: overexpression of hepatocyte growth factor gene reduced ischemic injury without cerebral edema in rat model. Circulation 109:424–431

Villoslada P, Hauser SL, Bartke I, Unger J, Heald N, Rosenberg D, et al. (2000) Human nerve growth factor protects common marmosets against autoimmune encephalomyelitis by switching the balance of T helper cell type 1 and 2 cytokines within the central nervous system. J Exp Med 191:1799–1806

Weimann JM, Charlton CA, Brazelton TR, Hackman RC, Blau HM (2003) Contribution of transplanted bone marrow cells to Purkinje neurons in human adult brains. Proc Natl Acad Sci 100:2088–2093

Yanai J, Doetchman T, Laufer N, Maslaton J, Mor-Yosef S, Safran A, et al. (1995) Embryonic cultures but not embryos transplanted to the mouse's brain grow rapidly without immunosuppression. Int J Neurosci 81:21–26

Zhang SC, Wernig M, Duncan ID, Brustle O, Thomson JA (2001) In vitro differentiation of transplantable neural precursors from human embryonic stem cells. Nat Biotechnol 19:1129–1133

9 *Novel Gene Therapeutic Strategies for Neurodegenerative Diseases*

K.A. Maguire-Zeiss, H.J. Federoff

9.1 Introduction

The treatment of patients with neurodegenerative disease poses unique challenges for clinicians. In particular, degenerative disorders of the central nervous system are vexing as the etiologies of many are inadequately defined. Therefore, almost all treatments focus on the relief of symptoms and not the prevention of disease. Herein we will focus on the treatment of one such disorder, Parkinson's disease. Following a discussion of the clinical manifestations, hallmark pathology, and current accepted treatments for this illness we will discuss potential novel therapies.

9.2 Clinical and Pathological Manifestations

Parkinson's disease (PD) affects approximately one million patients in the United States, with 40 % of those under the age of 60 years. This age-related progressive neurodegenerative disorder was first described in 1817 by James Parkinson in the classic "Essay on the Shaking Palsy" (Parkinson 1817). Clinically, PD patients present with symptoms which are typified by motoric dysfunction (Fahn 2003). In particular, PD patients present with tremor at rest, rigidity, bradykinesia, loss of postural reflexes, and flexed posture. In addition, the freezing phenomenon, that is the inability to move their feet forward, is often present. Although all six cardinal features are not required for a definitive diagnosis of PD, both rest tremor and bradykinesia are requisite. Eventually PD prevents patients from attending to the activities of daily living. Less appreciated are the nonmotor complications which can include depression and dementia, the former occurring more often in patients presenting at an earlier age (Kurlan 2003).

Parkinson's disease symptoms begin gradually but are progressive. Typically rest tremor or bradykinesia are the primary symptoms. These early symptoms are the result of the loss of nigrostriatal dopamine neurons and the attendant decrease in dopamine within the basal ganglia circuitry. As the disease progresses, other nonmotor symptoms develop which include changes in personality, behavior, cognition, sleep, and autonomic dysfunction (Benbunan et al. 2004). Indeed, loss of cardiac sympathetic innervation with the resultant orthostasis and postprandial light-headedness are common clinical features (Singleton et al. 2004). In addition to clinical indicators, neuroimaging aids in the definitive diagnosis of PD. Positron emission tomography (PET) and single-photon emission computed tomography (SPECT) imaging of PD patients reveals a marked decrease in dopaminergic nerve terminals within the striatum (Whone et al. 2003; Antonini and DeNotaris 2004; Eckert and Eidelberg 2004). The onset of clinical symptoms has been correlated with an apparent 80 % reduction of striatal dopamine reflecting the loss of nigrostriatal dopaminergic neurons.

In addition to clinical and neuroimaging features, the neuropathological hallmarks are specific loss of midbrain nigrostriatal projection neurons, gliosis, and the characteristic Lewy body, an intracytoplasmic inclusion found within the surviving dopaminergic neurons. These eosinophilic inclusions contain accumulated proteins, most notably ubiquitin and fibrillar α-synuclein (Baba et al. 1998). α-Synuclein, a previously unheralded presynaptic protein, moved to the forefront of PD etiopathogenesis with the description of several kindreds harboring either dominantly inherited mutations and or triplication of this wild-type gene (Polymeropoulos et al. 1996, 1997; Kitada et al. 1998; Kruger et al. 1998; Polymeropoulos 2000; Singleton et al. 2003).

9.3 Etiology and the Common Pathway Model

Familial mutations in α-synuclein account for only a small percentage of all PD cases (Benbunan et al. 2004; Scherfler et al. 2004). The remaining sporadic cases have an unknown etiology. Environmental insults have been implicated as a risk factor for PD (Seidler et al. 1996; Gorell et al. 1998; Mizuno et al. 1999; Olanow and Tatton 1999; Thiruchelvam et al. 2000a, b, 2002). However, environmental toxicants alone cannot account for the majority of sporadic PD cases. With this in mind, it has been proposed that the combination of genetic vulnerability and environmental agents act synergistically to initiate a cascade of cellular events leading to PD (Fig. 1; Maguire-Zeiss and Federoff 2003; Warner and Schapira 2003). Furthermore, α-synuclein is also posited to play a central role in disease pathogenesis in sporadic forms.

The invariant loss of nigrostriatal dopamine neurons further supports the idea that this disease shares a common pathobiologic course despite varying initiating events and requires a unique combination of factors present specifically within discrete and anatomically circumscribed regions. The centrality of α-synuclein underlies an unifying feature of the "common pathway" model. Accordingly, the consequence of various initiating events is presynaptic injury and nigrostriatal dysfunction which are clinically silent antecedents of posited irreversible damage. Cellular compensatory mechanisms are activated and these we surmise contribute to misfolding of proteins such as α-synuclein into conformation states that are cytotoxic. This pathologic folding may compromise proteosomal, lysosomal, and/or mitochondrial function. This progressive cell biologic injury unfolds slowly and results in bioenergetic failure,

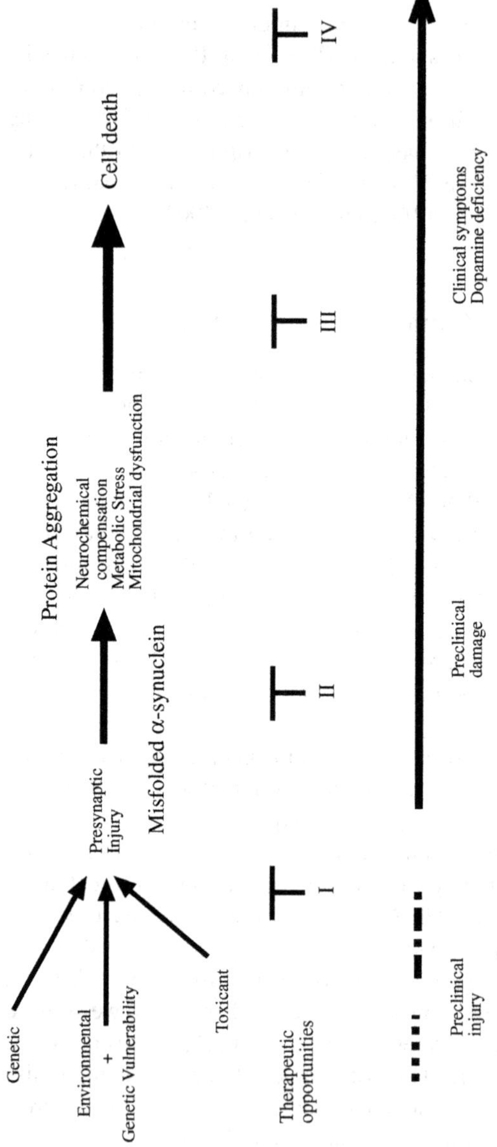

I: Specific triggering mechanisms
II: Shared early pathway step prior to presynaptic dopamine dysfunction; targeting misfolded α-synuclein
III: Shared later pathway step when dopaminergic neuron dysfunction occurs
IV: Restoring dopamine biosynthesis in denervated striatum

Fig. 1. Parkinson's disease common pathway model. (Reprinted by permission from The New York Academy of Sciences, Maguire-Zeiss and Federoff 2003)

the marked increase in reactive oxygen species generation and finally cellular demise. Therapies targeted to interdict either an initiating event, if elucidated, or at a common node in the pathway represent viable therapeutic strategies.

9.4 Parkinson's Disease Therapy

Parkinson's disease therapy is currently limited to treating symptoms. If the mechanisms underlying disease production were understood this may enable the recognition of preclinical or very early stage disease and consequently provide new therapeutic opportunities. As more information becomes available, improved therapy will ensue. Currently, PD treatment can be classified into three groups: pharmacological, surgical, and gene therapeutic.

9.4.1 Pharmacological

Dopaminergic Agents

This small molecule therapy is directed towards relieving motor dysfunction by controlling tremor, movement, and balance. Currently, the first line of treatment is oral administration of levodopa (L-DOPA), a precursor of dopamine, which serves to increase available dopamine in the neurotransmitter depleted denervated striatum (Fahn 2003; Nutt 2003; Stocchi 2003). This requires intact and functional postsynaptic neurons to be effective. In general, L-DOPA treatment is successful in reversing the characteristic akinesia and bradykinesia (Cotzias et al. 1967; Nutt 2003). However, over time this therapy becomes ineffective, and following the L-DOPA "honeymoon" period, patients develop levodopa-induced dyskinesia that eventually becomes intolerable and very limiting (Ahlskog and Muenter 2001). The on–off phenomenon often encountered following long-term L-DOPA treatment detracts from the drug's benefits and despite adjunctive therapy is not readily controlled. Currently, clinical interest lies in adjusting dosages and adding novel adjunctive therapy to prevent the development of the on–off phenomenon. Strides in this area have been hampered by a lack of drug response studies in animal models which recapitulate the on–off phenomenon. Recently, however, Lee et al. utilized a rat model to study the response of forepaw adjusting step following chronic small doses of L-DOPA (Lee et al. 2003). This model more faithfully reflects limb bradykinesia seen in patients and may prove useful in future studies of dosing and adjunct therapy.

Other pharmacological dopaminergic agents are also utilized early in the course of disease (reviewed in Fahn 2003). These include dopamine agonists, dopamine "releasers", and inhibitors of peripheral decarboxylase, catechol-O-methyl transferase, and dopamine metabolism. For example, carbidopa and benserazide are given in conjunction with L-DOPA, inhibiting decarboxylase activity and therefore the conversion to dopamine in the periphery. Entacapone, a catechol-O-methyl transferase inhibitor, is also utilized as an adjunct therapy preventing L-DOPA catabolism. Selegiline and rasagiline are examples of MAO type B inhibitors which extend the half-life of dopamine. Dopamine agonists such as bromocriptine, cabergoline, and apomorphine are the second most utilized class of PD drugs. In fact, Olanow et al. suggest that the revised PD treatment algorithm should include dopamine agonists as a first-line therapy for newly diagnosed PD patients because agonists reduce the risk of dyskinesia when compared to L-DOPA (Olanow et al. 2001).

Newer agents such as adenosine A2A antagonists also counteract the motor impairment typically seen in rat models of PD (Morelli 2003). Adenosine A2A receptors are expressed on striatal medium-spiny neurons and regulate striatal outflow to basal ganglia structures which control motor behavior (Fink et al. 1992). Under physiological conditions, adenosine binds these G protein-coupled receptors stimulating adenylyl cyclase and acting in opposition to dopamine D2 receptors (Svenningsson et al. 1997; Sebastiao and Ribeiro 2000). The adenosine A2A receptors antagonists, KW-6002 and caffeine, block this interaction and improve motor dysfunction (Chen et al. 2001; Bibbiani et al. 2003).

Antioxidants

Oxidative damage is proposed to play a role in Parkinson's disease pathogenesis. Likewise, mitochondrial dysfunction has been implicated in PD and is supported by studies utilizing the neurotoxicant, MPTP, a mitochondrial complex I inhibitor. Exposure of mice to MPTP results in progressive degeneration of nigrostriatal neurons with the concomitant appearance of behavioral dysfunction consistent with PD (Przedborski et al. 2001). In a separate study, α-synuclein transgenic mice treated with MPTP demonstrated extensive mitochondrial alterations, including an increase in mitochondrial size (Song et al. 2004). Transmission electron microscopy showed filamentous neuritic aggregations, axonal degeneration, and formation of electron dense perinuclear cytoplasmic inclusions specific to the substania nigra. Together these data suggest that an inhibitor of complex I activity can result in restricted nigrostriatal pathology supporting the role of oxidative injury and mitochondrial dysfunction in PD. Furthermore, α-synuclein overexpression increases the vulnerability of dopaminergic neurons to mitochondrial complex I inhibitors. Therapies designed to improve mitochondrial function may therefore prove useful in the treatment of PD.

Several agents are available with demonstrated antioxidant effects in animal models that may prove valuable in the treatment of human disease. These include ubiquinone/coenzyme Q_{10} (CoQ_{10}) and creatine (reviewed in Beal 2003). Creatine was shown to be protective in a mouse model of PD. MPTP-treated mice receiving creatine demonstrated a reduction in dopaminergic neuron loss as well as an increase in dopamine content (Matthews et al. 1998; Beal 2003). Currently, creatine is being evaluated in clinical trials for PD (Tarnopolsky and Beal 2001). Recently CoQ_{10} has received much public attention. This antioxidant scavenges free radicals in the inner mitochondrial membrane (Kagan et al. 1990). CoQ_{10} also interacts with mitochondrial uncoupling proteins facilitating free radical reduction (Casteilla et al. 2001; Echtay et al. 2001). CoQ_{10} has been utilized in clinical trials of patients with several neurodegenerative diseases (Koroshetz et al. 1997; Huntington's Study Group 2001; Shults et al. 2002). Oral administration of CoQ_{10} was well-tolerated, resulted in dose-dependent plasma levels and resulted in a slowing of disease progression (Huntington's Study Group 2001; Shults et al. 2002). In the PD clinical trial, there was evidence of slowing of the functional decline and a linear trend between dosage and mean change in the total Unified Parkinson's Disease Rating Scale (UPDRS). Furthermore, a comparison of treatment group with placebo group demonstrated a significant difference in the UPDRS (Shults et al. 2002). The PD clinical trial results portend a promising future for CoQ_{10}.

Neuroprotective Agents

The dopaminergic and antioxidant agents discussed above alleviate PD symptoms but are not curative. Neuroprotection or neurorestoration of nigrostriatal neurons is the most lauded goal of PD therapy. Strategies have largely centered on delivery of potent neurotrophic factors either directly or through cellular replacement or viral vector gene therapy strategies. The molecule that has moved to the forefront of PD neuroprotective therapy is glial cell line- derived neurotrophic factor (GDNF). GDNF was discovered to be dopaminotrophic, involved in dopamine neuron survival, and reduced in human PD basal ganglia, making it the logical choice for further study (Lin et al. 1993; Chauhan et al. 2001; Hurelbrink and Barker 2001). Initial strategies focused on direct injection of GDNF in rodent models of PD. A number of laboratories demonstrated both protective and restorative effects of GDNF, including improvements in behavioral deficits, dopaminergic fiber outgrowth and neuronal survival (Opacka-Juffry et al. 1995; Tomac et al. 1995; Aoi et al. 2000, 2001; Kirik et al. 2000a,b). Intraventricular administration of GDNF is also neuroprotective in MPTP-treated nonhuman primates (Gash et al. 1996; Gerhardt et al. 1999; Costa et al. 2001; Iravani et al. 2001; Grondin et al. 2002). Both rodent and nonhuman primate models of parkinsonism have been treated with GDNF via cell-based and viral-based gene therapy delivery methods with similar success (Lindner et al. 1995; Choi-Lundberg et al. 1997; Kordower et al. 2000; Nakao et al. 2000; Akerud et al. 2001; Date et al. 2001; Cunningham and Su 2002; Vigna et al. 2002; Hurelbrink and Barker 2004).

The first clinical trial of intraventricular GDNF was initiated by Amgen Inc. (Thousand Oaks, CA) but failed due to treatment-related side effects and lack of efficacy. Rather than reflecting a lack of concordance between animal models and clinical outcomes, Kordower and others have demonstrated that the lack of efficacy was due to the neurotrophin delivery method (Kordower 2003). Kordower et al., utilizing site-specific gene delivery of GDNF to three different model systems (normal, aged, and MPTP-treated monkeys), have demonstrated safety and efficacy (Kordower 2003). A second clinical trial employed direct injection of GDNF via putaminal implantation of a catheter and pump into five idiopathic L-DOPA-responsive PD patients (Gill et al. 2003). Utilizing the UPDRS, 1 year later, there was a 39 % improvement in the off-medication motor subscore and medication-induced dyskinesias were reduced by 64 %. Furthermore, PET scans demonstrated a 28 % increase in putaminal dopamine storage after 18 months. Finally, there were no GDNF-related serious clinical side-effects. Clearly these studies have paved the way for continued examination of GDNF as a potential PD therapy. The current GDNF Phase II clinical trial has not come to completion but the PD community is anxiously awaiting the results.

In addition to traditional neurotrophic factors, dopamine agonists are also being studied as potential neuroprotective agents. Preclinical studies demonstrate that treatment of cultured neurons with dopamine agonists protects against toxicant effects and enhances neuron growth (Carvey et al. 1997; Gassen et al. 1998; Kitamura et al. 1998; Takashima et al. 1999; Zou et al. 1999; Schapira and Olanow 2003). Animal models of parkinsonism respond to agonist treatment with amelioration of symptoms and protection of substantia nigra dopaminergic neurons (Ogawa et al. 1994; Asanuma et al. 1995; Kitamura et al. 1997; Muralikrishnan and Mohanakumar 1998; Grunblatt et al. 1999; Vu et al. 2000; Jenner 2002; Schapira and Olanow 2003). Taken together these studies suggest that dopamine agonists may not only be effective for symptomatic treatment of PD but could also be neuroprotective therapy. Greater therapeutic benefit may be achieved when neuroimaging and genomic technology allow for earlier diagnosis and thus earlier intervention for PD patients.

9.4.2 Surgical

Subsets of PD patients refractory to pharmacological therapy or exhibiting uncontrollable side effects are candidates for surgical treatment. As with the pharmacological therapies, surgical treatments provide symptomatic control and include ablation or deep brain stimulation (DBS; Fink et al. 1992; Benabid et al. 1998; Krack et al. 1998; Benabid 2003; Hamani and Lozano 2003; Kalia et al. 2004). As new techniques in both neuroimaging and surgery evolve, this intervention will likely become more available. The goal of surgical therapy is to suppress the abnormal neural activity in the basal ganglia, relieving the motor dysfunction. Currently ablative lesioning of subthalamic nuclei carries a greater risk of concomitant neurological deficits than DBS. DBS is designed to provide a controlled electrical signal to silence an excessively active brain nucleus, thus alleviating symptoms. The advantage of DBS over ablative lesions is that the stimulation can be precisely regulated and if needed discontinued. Importantly, DBS affords clinicians the opportunity to explore putative targets intraoperatively. In fact, DBS is more precise than pharmacological interventions since the electrodes are localized to specific nuclei. Although DBS mimics lesions, unlike ablative therapy DBS is reversible, leaving open the option for other therapies as they become available. A downside of DBS is that it requires a state-of-the-art surgical center and is expensive. However, DBS is a functional neurosurgical tool that when added to the PD armamentarium can prolong the symptom-free phase.

9.4.3 Novel Therapies

The goals of PD therapy are to prevent disease, protect neurons, and restore function, a tall order for a disease with an unknown etiology. Therapy can be divided into several potential approaches (Fig. 1). First, therapy can be targeted to the specific disease mechanism (I). Mechanism-directed therapy would employ strategies to "knock-down" activity of a toxic gene product through a loss of function (LOF) approach or provide expression of a prosurvival gene production through a gain of function (GOF). Selection of a particular strategy is driven by an explicit understanding of the operative pathophysiologic mechanism(s). For example, autosomal recessive inherited juvenile parkinsonism (ARJP), owing to loss of parkin function, might be amenable to restoration of expression of the mutated gene. Similarly, to mitigate the long term consequences of antecedent toxicant exposure with sustained cytotoxic actions, therapeutic therapies could be directed towards detoxification. Finally, the combined action of genetic and possible environmental exposure, suggested in kindreds harboring Nurr1 gene polymorphisms, potentially could benefit from a gene therapy designed to augment Nurr1 action.

Second, therapy can be targeted to the presynaptic injury phase of the shared common pathway model (II). Gene therapy approaches would be designed to either target transgene expression to the degenerating nigrostriatal neuronal population or their striatal projection fields. Such strategies could employ therapeutic transgenes encoding neurotrophic factors, antiapoptotic proteins, antioxidant proteins, or gene products acting to enhance the refolding or degradation of presumptive toxic misfolded proteins.

Third, therapy can be directed toward a later shared pathway step occurring downstream of dopaminergic neuron dysfunction in an effort to save the remaining neurons (III). Here, therapeutic transgenes would include neuroprotective and neurorestorative growth factors and possibly antiapoptotic molecules.

Finally, therapy can be targeted toward the restoration of dopamine biosynthesis (IV) in the denervated striatum via a delivery of dopamine biosynthesis enzymes such as amino acid decarboxylase (AADC), tyrosine hydroxylase, and GTP cyclohydrolase. These therapies are not mutually exclusive, and are envisioned to largely depend upon the individual patient and disease state. With each therapy the important objective is to develop the necessary technology to efficiently and specifically deliver the therapeutic agent to the target cell.

Delivery of Therapeutic Genes

The delivery of therapeutic genes to the compromised region(s) is a common goal of PD therapy. Once the target gene and cell are identified, delivery approaches need to be considered. Gene delivery methods include direct delivery of naked or encapsulated DNA and delivery of a vector system via cell-based or viral-based therapy. Progress on different vector approaches has been made but only those in or about to enter clinical trials will be highlighted.

Cell-based therapies are being explored for PD. Initially, fetal tissue transplantation had been utilized in a limited number of PD cases with variable success (Freed et al. 2001). A Phase I clinical trial for Alzheimer's disease utilizing autologous fibroblasts genetically engineered to produce nerve growth factor has been completed. A similar methodology could be employed to provide GDNF in PD. Other methodologies include the use of stem cells and astrocytes as delivery vehicles. Cell-based therapy is limited by the requirement to produce a population of gene-modified cells in sufficient numbers, their production of other secreted gene products, and the spread of these cells following implantation. Many of these shortcomings may be overcome through direct viral-based gene therapy.

Targeting Restoration of Dopamine Biosynthesis (IV)

The first viral-based gene therapy trial for Parkinson's disease was begun on August 18, 2003. In this study, supported by Neurologix, Inc. (Fort Lee, NJ) adeno-associated virus containing the glutamic acid decarboxylase gene isoforms 65 kDa and 67 kDa (GAD) was infused into the subthalamic nucleus of a 55-year-old PD patient (Howard 2003; Oransky 2003). GAD is responsible for the synthesis of the inhibitory neurotransmitter, γ-aminobutyric acid (GABA). The goal of this symptomatic therapy is to phenoconvert the hyperexcitable glutamatergic subthalamic nucleus to a GABAergic inhibitory nucleus, thereby silencing its pathologic output. Currently no data are available regarding the efficacy of this initial treatment. To date four patients have been enrolled and have undergone the neurosurgically delivered gene therapy.

Targeting Dopaminergic Neuron Dysfunction (II/III)

A plethora of in vivo GDNF gene delivery data in rodent and nonhuman primate models of PD has vaulted this neurotrophin to the forefront of PD neuroprotective therapy (Kordower 2003). The effectiveness of GDNF will likely depend on the number of neurons still present in the diseased brain at the time of therapy. Such a consideration would suggest patients with earlier stage disease may be most amenable for this form of therapy. Furthermore, this therapy may not be limited to GDNF as other GDNF family ligands (GFLs) such as neurturin, persephin, and artemin also support midbrain dopamine neurons. These GFLs also signal through c-ret and bind to GFRα coreceptors (Akerud et al. 2002; Marco et al. 2002; Sariola and Saarma 2003).

Prior to initiating clinical trials of GDNF gene therapy, the most effective and apparently safe delivery method should be established. Viral-based delivery methods appear most promising and some have been documented in rodents and nonhuman primates (Kirik et al. 2000a,b; Kordower et al. 2000; Bowers et al. 2001; Freed et al. 2001; Hsich et al. 2002; Glorioso et al. 2003). Although a number viral vectors are available, the most currently viable candidates for PD gene therapy are adeno-associated and lentivirus virus vectors because they confer long-term gene expression and have limited known side-effects (Kingsman 2003; Mochizuki and Mizuno 2003). Moreover, these two vectors are intrinsically neurotrophic and thus capable of targeting diseased neurons.

Targeting Shared Early Pathway Steps (II)

In the preclinical phase of PD we postulate the existence of a shared biochemical node common to all forms of parkinsonism and which may represent a target for gene therapy. In our model we hypothesize that the misfolding of α-synuclein to a toxic conformer is among the most critical events in the pathogenic pathway (Fig. 1; II). Preventing the formation of this proposed toxic α-synuclein conformer may be a therapeutically promising approach. Our group is pursuing a gene therapy approach to interfere with α-synuclein misfolding. The mechanism(s) driving misfolding of α-synuclein in dopaminergic neurons remains an active area of current investigation.

The normal role of α-synuclein in dopaminergic neurons is unknown. This 14 kDa cytosolic protein is abundant in presynaptic terminals and normally adopts a random-coil structure (Maroteaux et al. 1988; Jakes et al. 1994; Iwai et al. 1995; Clayton and George 1998; Rochet et al. 2004). Cell culture and animal models of either α-synuclein overexpression or deletion suggest that it plays an important role in synaptic function and plasticity (Clayton and George 1998; Hsu et al. 1998; Takenouchi et al. 2001). It may function as a molecular chaperone through interactions with other proteins including synphilin, parkin, tyrosine hydroxylase, phospholipase D, and α-synuclein (Ischiropoulos 2003). Interaction between α-synuclein and the dopamine transporter (DAT) modulates transporter localization to the synaptic membrane, suggesting a role in dopamine recycling (Lee et al. 2001).

The changed secondary structure of α-synuclein may represent the pathological event in dopaminergic neurons. Progression results in the formation of intracytoplasmic Lewy bodies containing fibrillar α-synuclein. Normal α-synuclein is degraded by the proteosome (Weinreb et al. 1996). Upon interaction with synthetic lipid bilayers, α-synuclein assumes an amphipathic α-helical structure (Davidson et al. 1998; Perrin et al. 2000; Chandra et al. 2003). In the face of oxidative stress engendered by neurotoxicant exposure or changes in temperature, pH, molecular crowding or agitation, there is a transformation from random coil conformation to a β-sheet structure that forms protofibrils. These seed further aggregation to produce fibrils which are resistant to proteosomal degradation and accumulate within the cytoplasm. Dopamine drives the formation of abnormal α-synuclein conformations (Asanuma et al. 2003; Rochet et al. 2004). Inherently reactive, dopamine is readily oxidized to a quinone during which superoxide and hydrogen peroxide are released. Dopamine quinone adducts form on α-synuclein in vitro and these slow protofibril-to-fibril conversion (Conway et al. 2001). Covalent dopamine:synuclein adducts have yet to be identified in vivo (cells or tissue), but such evidence would provide a clear link between oxidative stress, dopamine quinone formation, and α-synuclein misfolding in PD pathogenesis.

In dopamine neurons, the native α-synuclein (random coil), the protofibrillar conformer, and the aggregated form are postulated to be in steady state. In this scenario oxidative stress typified by the formation of dopamine quinone will drive the native α-synuclein conformers towards toxic oligomers. As a molecular instigator of PD, α-synuclein may be analogous to amyloid beta (Aβ) which forms extracellular deposits and is a pathological hallmark of Alzheimer's disease (Ogomori et al. 1989; Mann et al. 1990; Suenaga et al. 1990; Forloni et al. 2002). We suggest that therapies specifically designed to interact and reduce toxic α-synuclein conformers will attenuate disease progression. Thus, preventing the formation of toxic protofibrillar forms of α-synuclein should be a goal of future therapeutics.

Our laboratory and others are pursuing such a gene therapy approach whereby conformation-specific, single-chain antibodies (scFvs) are expressed and intended to interfere with α-synuclein protofibrillar formation (Emadi et al. 2004). Single-chain antibodies are composed of the minimal antigen binding site formed by noncovalent association of the V_H and V_L variable domains joined by a flexible polypeptide linker. Human scFv-phage libraries are available and allow for high affinity human scFv antibodies to be selected from combinatorial libraries (Malone and Sullivan 1996). Further antibody engineering makes it possible to manipulate the genes encoding the scFvs, allowing for antibody binding site expression within mammalian cells (intrabodies). A similar approach has been utilized for Huntington's disease where human huntingtin-specific scFv intrabodies disrupted huntingtin aggregation (Lecerf et al. 2001; Murphy and Messer 2004).

In PD we suggest that protein aggregates forming within dopamine neurons are seeded, require the protofibrillar conformers of α-synuclein (β-sheet), and are cytotoxic. Identification of scFvs that will selectively recognize the misfolded or protofibrillar conformers of α-synuclein is anticipated to attenuate protein toxicity. In this scenario, scFvs interact with α-synuclein prior to aggregation, thus depleting the pool of available protein. Alternatively, scFvs bind to and deplete protofibrillar α-synuclein, thus attenuating further aggregation. A third possibility is that scFvs interact with aggregated α-synuclein and efficiently target the complex for degradation. Any of these proposed strategies would decrease protein toxicity and potentially mitigate damage. Once identified, conformer-specific scFv could be expressed within the brain utilizing the most efficient viral-based gene therapy approach.

Other molecular therapy approaches are possible and would include the use of small molecules to interfere with protein activity. Conway et al. demonstrated that drug-like molecules inhibited α-synuclein fibrilization (Conway et al. 2001). This group identified 15 fibril inhibitors, 14 of which were catecholamines related to dopamine. The finding that both dopamine and L-dopa inhibit fibril formation of α-synuclein has also been confirmed by another laboratory (Conway et al. 2001; Li et al. 2004). In the latter study, Aβ fibrils were also dissolved following catecholamine treatment, suggesting a wide range of therapeutic effects are possible with these small molecules.

9.5 Summary

The convergent pathobiologic model of Parkinson's disease stipulates that disparate insults initiate a disease process that obligately share a common pathway leading to cell death. A combinatorial treatment which targets various steps in this pathway is likely to be the most successful therapeutic strategy. As advances are made in the field of neuroimaging and pharmacogenomics, early detection of sporadic PD will become a reality. Early intervention will likely spare more dopaminergic neurons and extend the quality of life for the patient. Continued advancements in the fields of pharmacology, neurosurgery, and gene therapy will strengthen the armamentarium available for the treatment of PD patients.

Acknowledgements. This work was supported by DAMD17–03–1-0009 to HJF.

References

Ahlskog JE, Muenter MD (2001) Frequency of levodopa-related dyskinesias and motor fluctuations as estimated from the cumulative literature. Mov Disord 16:448–458

Akerud P, Canals JM, Snyder EY, Arenas E (2001) Neuroprotection through delivery of glial cell line-derived neurotrophic factor by neural stem cells in a mouse model of Parkinson's disease. J Neurosci 21:8108–8118

Akerud P, Holm PC, Castelo-Branco G, Sousa K, Rodriguez FJ, Arenas E (2002) Persephin-overexpressing neural stem cells regulate the function of nigral dopaminergic neurons and prevent their degeneration in a model of Parkinson's disease. Mol Cell Neurosci 21:205–222

Antonini A, DeNotaris R (2004) PET and SPECT functional imaging in Parkinson's disease. Sleep Med 5:201–206

Aoi M, Date I, Tomita S, Ohmoto T (2000) GDNF induces recovery of the nigrostriatal dopaminergic system in the rat brain following intracerebroventricular or intraparenchymal administration. Acta Neurochir (Wien) 142:805–810

Aoi M, Date I, Tomita S, Ohmoto T (2001) Single administration of GDNF into the striatum induced protection and repair of the nigrostriatal dopaminergic system in the intrastriatal 6-hydroxydopamine injection model of hemiparkinsonism. Restor Neurol Neurosci 17:31–38

Asanuma M, Ogawa N, Nishibayashi S, Kawai M, Kondo Y, Iwata E (1995) Protective effects of pergolide on dopamine levels in the 6- hydroxydopamine-lesioned mouse brain. Arch Int Pharmacodyn Ther 329:221– 230

Asanuma M, Miyazaki I, Ogawa N (2003) Dopamine- or L-DOPA- induced neurotoxicity: the role of dopamine quinone formation and tyrosinase in a model of Parkinson's disease. Neurotox Res 5:165–176

Baba M, Nakajo S, Tu PH, Tomita T, Nakaya K, Lee VM, Trojanowski JQ, Iwatsubo T (1998) Aggregation of alpha-synuclein in Lewy bodies of sporadic Parkinson's disease and dementia with Lewy bodies. Am J Pathol 152:879–884

Beal MF (2003) Bioenergetic approaches for neuroprotection in Parkinson's disease. Ann Neurol 53 Suppl 3: S39–47; discussion S47–48

Benabid AL (2003) Deep brain stimulation for Parkinson's disease. Curr Opin Neurobiol 13:696–706

Benabid AL, Benazzouz A, Hoffmann D, Limousin P, Krack P, Pollak P (1998) Long-term electrical inhibition of deep brain targets in movement disorders. Mov Disord 13 Suppl 3:119–125

Benbunan BR, Korczyn AD, Giladi N (2004) Parkin mutation associated parkinsonism and cognitive decline, comparison to early onset Parkinson's disease. J Neural Transm 111:47–57

Bibbiani F, Oh JD, Petzer JP, Castagnoli N, Jr., Chen JF, Schwarzschild MA, Chase TN (2003) A2A antagonist prevents dopamine agonist- induced motor complications in animal models of Parkinson's disease. Exp Neurol 184:285–294

Bowers WJ, Maguire-Zeiss KA, Harvey BK, Federoff HJ (2001) Gene therapeutic approaches to the treatment of Parkinson's disease. Clin Neurosci Res 1:483–495

Carvey PM, Pieri S, Ling ZD (1997) Attenuation of levodopa-induced toxicity in mesencephalic cultures by pramipexole. J Neural Transm 104:209–228

Casteilla L, Rigoulet M, Penicaud L (2001) Mitochondrial ROS metabolism: modulation by uncoupling proteins. IUBMB Life 52:181–188

Chandra S, Chen X, Rizo J, Jahn R, Sudhof TC (2003) A broken alpha-helix in folded alpha-Synuclein. J Biol Chem 278:15313–15318

Chauhan NB, Siegel GJ, Lee JM (2001) Depletion of glial cell line- derived neurotrophic factor in substantia nigra neurons of Parkinson's disease brain. J Chem Neuroanat 21:277–288

Chen JF, Moratalla R, Impagnatiello F, Grandy DK, Cuellar B, Rubinstein M, Beilstein MA, Hackett E, Fink JS, Low MJ, Ongini E, Schwarzschild MA (2001) The role of the D(2) dopamine receptor (D(2)R) in A(2A) adenosine receptor (A(2A)R)-mediated behavioral and cellular responses as revealed by A(2A) and D(2) receptor knockout mice. Proc Natl Acad Sci 98:1970–5

Choi-Lundberg DL, Lin Q, Chang YN, Chiang YL, Hay CM, Mohajeri H, Davidson BL, Bohn MC (1997) Dopaminergic neurons protected from degeneration by GDNF gene therapy. Science 275:838–841

Clayton DF, George JM (1998) The synucleins: a family of proteins involved in synaptic function, plasticity, neurodegeneration and disease. Trends Neurosci 21:249–254

Conway KA, Rochet JC, Bieganski RM, Lansbury PT, Jr. (2001) Kinetic stabilization of the alpha-synuclein protofibril by a dopamine-alpha- synuclein adduct. Science 294:1346–1349

Costa S, Iravani MM, Pearce RK, Jenner P (2001) Glial cell line- derived neurotrophic factor concentration dependently improves disability and motor activity in MPTP-treated common marmosets. Eur J Pharmacol 412:45–50

Cotzias GC, Van Woert MH, Schiffer LM (1967) Aromatic amino acids and modification of parkinsonism. N Engl J Med 276:374–379

Cunningham LA, Su C (2002) Astrocyte delivery of glial cell line- derived neurotrophic factor in a mouse model of Parkinson's disease. Exp Neurol 174:230–242

Date I, Shingo T, Yoshida H, Fujiwara K, Kobayashi K, Takeuchi A, Ohmoto T (2001) Grafting of encapsulated genetically modified cells secreting GDNF into the striatum of parkinsonian model rats. Cell Transplant 10:397–401

Davidson WS, Jonas A, Clayton DF, George JM (1998) Stabilization of alpha-synuclein secondary structure upon binding to synthetic membranes. J Biol Chem 273:9443–9449

Echtay KS, Winkler E, Frischmuth K, Klingenberg M (2001) Uncoupling proteins 2 and 3 are highly active H(+) transporters and highly nucleotide sensitive when activated by coenzyme Q (ubiquinone). Proc Natl Acad Sci 98:1416–1421

Eckert T, Eidelberg D (2004) The role of functional neuroimaging in the differential diagnosis of idiopathic Parkinson's disease and multiple system atrophy. Clin Auton Res 14:84–91

Emadi S, Liu R, Yuan B, Schulz P, McAllister C, Lyubchenko Y, Messer A, Sierks MR (2004) Inhibiting aggregation of alpha-synuclein with human single chain antibody fragments. Biochemistry 43:2871–2878

Fahn S (2003) Description of Parkinson's disease as a clinical syndrome. Ann N Y Acad Sci 991:1–14

Fink JS, Weaver DR, Rivkees SA, Peterfreund RA, Pollack AE, Adler EM, Reppert SM (1992) Molecular cloning of the rat A2 adenosine receptor: selective co-expression with D2 dopamine receptors in rat striatum. Brain Res Mol Brain Res 14:186–195

Forloni G, Terreni L, Bertani I, Fogliarino S, Invernizzi R, Assini A, Ribizzi G, Negro A, Calabrese E, Volonte MA, Mariani C, Franceschi M, Tabaton M, Bertoli A (2002) Protein misfolding in Alzheimer's and Parkinson's disease: genetics and molecular mechanisms. Neurobiol Aging 23:957

Freed CR, Greene PE, Breeze RE, Tsai WY, DuMouchel W, Kao R, Dillon S, Winfield H, Culver S, Trojanowski JQ, Eidelberg D, Fahn S (2001) Transplantation of embryonic dopamine neurons for severe Parkinson's disease. N Engl J Med 344:710–719

Gash DM, Zhang Z, Ovadia A, Cass WA, Yi A, Simmerman L, Russell D, Martin D, Lapchak PA, Collins F, Hoffer BJ, Gerhardt GA (1996) Functional recovery in parkinsonian monkeys treated with GDNF. Nature 380:252–255

Gassen M, Gross A, Youdim MB (1998) Apomorphine enantiomers protect cultured pheochromocytoma (PC12) cells from oxidative stress induced by H2O2 and 6-hydroxydopamine. Mov Disord 13:661–667

Gerhardt GA, Cass WA, Huettl P, Brock S, Zhang Z, Gash DM (1999) GDNF improves dopamine function in the substantia nigra but not the putamen of unilateral MPTP-lesioned rhesus monkeys. Brain Res 817:163–171

Gill SS, Patel NK, Hotton GR, O'Sullivan K, McCarter R, Bunnage M, Brooks DJ, Svendsen CN, Heywood P (2003) Direct brain infusion of glial cell line-derived neurotrophic factor in Parkinson disease. Nat Med 9:589–595

Glorioso JC, Mata M, Fink DJ (2003) Therapeutic gene transfer to the nervous system using viral vectors. J Neurovirol 9:165–172

Gorell JM, Johnson CC, Rybicki BA, Peterson EL, Richardson RJ (1998) The risk of Parkinson's disease with exposure to pesticides, farming, well water, and rural living. Neurology 50:1346–1350

Grondin R, Zhang Z, Yi A, Cass WA, Maswood N, Andersen AH, Elsberry DD, Klein MC, Gerhardt GA, Gash DM (2002) Chronic, controlled GDNF infusion promotes structural and functional recovery in advanced parkinsonian monkeys. Brain 125:2191–2201

Grunblatt E, Mandel S, Berkuzki T, Youdim MB (1999) Apomorphine protects against MPTP-induced neurotoxicity in mice. Mov Disord 14:612–618

Hamani C, Lozano AM (2003) Physiology and pathophysiology of Parkinson's disease. Ann N Y Acad Sci 991:15–21

Howard K (2003) First Parkinson gene therapy trial launches. Nat Biotechnol 21:1117–1118

Hsich G, Sena-Esteves M, Breakefield XO (2002) Critical issues in gene therapy for neurologic disease. Hum Gene Ther 13:579–604

Hsu LJ, Mallory M, Xia Y, Veinbergs I, Hashimoto M, Yoshimoto M, Thal LJ, Saitoh T, Masliah E (1998) Expression pattern of synucleins (non- Abeta component of Alzheimer's disease amyloid precursor protein/alpha- synuclein) during murine brain development. J Neurochem 71:338–344

Huntington's Study Group, T (2001) A randomized, placebo- controlled trial of coenzyme Q_{10}, remacemide in Huntington's disease. Neurology 57:397–404

Hurelbrink CB, Barker RA (2001) Prospects for the treatment of Parkinson's disease using neurotrophic factors. Expert Opin Pharmacother 2:1531–1543

Hurelbrink CB, Barker RA (2004) The potential of GDNF as a treatment for Parkinson's disease. Exp Neurol 185:1–6

Iravani MM, Costa S, Jackson MJ, Tel BC, Cannizzaro C, Pearce RK, Jenner P (2001) GDNF reverses priming for dyskinesia in MPTP-treated, L- DOPA-primed common marmosets. Eur J Neurosci 13:597–608

Ischiropoulos H (2003) Oxidative modifications of alpha-synuclein. Ann N Y Acad Sci 991:93–100

Iwai A, Masliah E, Yoshimoto M, Ge N, Flanagan L, de Silva HA, Kittel A, Saitoh T (1995) The precursor protein of non-A beta component of Alzheimer's disease amyloid is a presynaptic protein of the central nervous system. Neuron 14:467–475

Jakes R, Spillantini MG, Goedert M (1994) Identification of two distinct synucleins from human brain. FEBS Lett 345:27–32

Jenner P (2002) Pharmacology of dopamine agonists in the treatment of Parkinson's disease. Neurology 58: S1–8

Kagan V, Serbinova E, Packer L (1990) Antioxidant effects of ubiquinones in microsomes and mitochondria are mediated by tocopherol recycling. Biochem Biophys Res Commun 169:851–857

Kalia SK, Nash JE, Lozano AM (2004) To serve and protect? Interventions in the subthalamic nucleus for Parkinson's disease. Commentary on "Ablation of the subthalamic nucleus protects dopaminergic phenotype but not cell survival in a rat model of Parkinson's disease". Exp Neurol 185:201–203

Kingsman SM (2003) Lentivirus: a vector for nervous system applications. Ernst Schering Res Found Workshop: 179–207

Kirik D, Rosenblad C, Bjorklund A (2000a) Preservation of a functional nigrostriatal dopamine pathway by GDNF in the intrastriatal 6-OHDA lesion model depends on the site of administration of the trophic factor. Eur J Neurosci 12:3871–3882

Kirik D, Rosenblad C, Bjorklund A, Mandel RJ (2000b) Long-term rAAV-mediated gene transfer of GDNF in the rat Parkinson's model: intrastriatal but not intranigral transduction promotes functional regeneration in the lesioned nigrostriatal system. J Neurosci 20:4686–4700

Kitada T, Asakawa S, Hattori N, Matsumine H, Yamamura Y, Minoshima S, Yokochi M, Mizuno Y, Shimizu N (1998) Mutations in the parkin gene cause autosomal recessive juvenile parkinsonism. Nature 392:605–608

Kitamura Y, Kohno Y, Nakazawa M, Nomura Y (1997) Inhibitory effects of talipexole and pramipexole on MPTP-induced dopamine reduction in the striatum of C57BL/6 N mice. Jpn J Pharmacol 74:51–57

Kitamura Y, Kosaka T, Kakimura JI, Matsuoka Y, Kohno Y, Nomura Y, Taniguchi T (1998) Protective effects of the antiparkinsonian drugs talipexole and pramipexole against 1-methyl-4-phenylpyridinium-induced apoptotic death in human neuroblastoma SH-SY5Y cells. Mol Pharmacol 54:1046–1054

Kordower JH (2003) In vivo gene delivery of glial cell line–derived neurotrophic factor for Parkinson's disease. Ann Neurol 53 Suppl 3:S120–132; discussion S132–134

Kordower JH, Emborg ME, Bloch J, Ma SY, Chu Y, Leventhal L, McBride J, Chen EY, Palfi S, Roitberg BZ, Brown WD, Holden JE, Pyzalski R, Taylor MD, Carvey P, Ling Z, Trono D, Hantraye P, Deglon N, Aebischer P (2000) Neurodegeneration prevented by lentiviral vector delivery of GDNF in primate models of Parkinson's disease. Science 290:767–773

Koroshetz WJ, Jenkins BG, Rosen BR, Beal MF (1997) Energy metabolism defects in Huntington's disease and effects of coenzyme Q_{10}. Ann Neurol 41:160–165

Krack P, Benazzouz A, Pollak P, Limousin P, Piallat B, Hoffmann D, Xie J, Benabid AL (1998) Treatment of tremor in Parkinson's disease by subthalamic nucleus stimulation. Mov Disord 13:907–914

Kruger R, Kuhn W, Muller T, Woitalla D, Graeber M, Kosel S, Przuntek H, Epplen JT, Schols L, Riess O (1998) Ala30Pro mutation in the gene encoding alpha-synuclein in Parkinson's disease. Nat Genet 18:106–108

Kurlan R (2003) Declining medication requirement in some patients with advanced Parkinson disease and dementia. Clin Neuropharmacol 26:171

Lecerf JM, Shirley TL, Zhu Q, Kazantsev A, Amersdorfer P, Housman DE, Messer A, Huston JS (2001) Human single-chain Fv intrabodies counteract in situ huntingtin aggregation in cellular models of Huntington's disease. Proc Natl Acad Sci 98:4764–4769

Lee EA, Lee WY, Kim YS, Kang UJ (2003) The effects of chronic L- DOPA therapy on pharmacodynamic parameters in a rat model of motor response fluctuations. Exp Neurol 184:304-12

Lee FJ, Liu F, Pristupa ZB, Niznik HB (2001) Direct binding and functional coupling of alpha-synuclein to the dopamine transporters accelerate dopamine-induced apoptosis. FASEB J 15:916–926

Li J, Zhu M, Manning-Bog AB, Di Monte DA, Fink AL (2004) Dopamine and L-dopa disaggregate amyloid fibrils: implications for Parkinson's and Alzheimer's disease. FASEB J 18:962–964

Lin LF, Doherty DH, Lile JD, Bektesh S, Collins F (1993) GDNF: a glial cell line-derived neurotrophic factor for midbrain dopaminergic neurons. Science 260:1130–1132

Lindner MD, Winn SR, Baetge EE, Hammang JP, Gentile FT, Doherty E, McDermott PE, Frydel B, Ullman MD, Schallert T, et al. (1995) Implantation of encapsulated catecholamine and GDNF-producing cells in rats with unilateral dopamine depletions and parkinsonian symptoms. Exp Neurol 132:62–76

Maguire-Zeiss KA, Federoff HJ (2003) Convergent pathobiologic model of Parkinson's disease. Ann N Y Acad Sci 991:152–166

Malone J, Sullivan MA (1996) Analysis of antibody selection by phage display utilizing anti-phenobarbital antibodies. J Mol Recognit 9:738–745

Mann DM, Jones D, Prinja D, Purkiss MS (1990) The prevalence of amyloid (A4) protein deposits within the cerebral and cerebellar cortex in Down's syndrome and Alzheimer's disease. Acta Neuropathol (Berl) 80:318–327

Marco S, Canudas AM, Canals JM, Gavalda N, Perez-Navarro E, Alberch J (2002) Excitatory amino acids differentially regulate the expression of GDNF, neurturin, and their receptors in the adult rat striatum. Exp Neurol 174:243–252

Maroteaux L, Campanelli JT, Scheller RH (1988) Synuclein: a neuron-specific protein localized to the nucleus and presynaptic nerve terminal. J Neurosci 8:2804–2815

Matthews RT, Yang L, Jenkins BG, Ferrante RJ, Rosen BR, Kaddurah-Daouk R, Beal MF (1998) Neuroprotective effects of creatine and cyclocreatine in animal models of Huntington's disease. J Neurosci 18:156–163

Mizuno Y, Shimoda-Matsubayashi S, Matsumine H, Morikawa N, Hattori N, Kondo T (1999) Genetic and environmental factors in the pathogenesis of Parkinson's disease. Adv Neurol 80:171–179

Mochizuki H, Mizuno Y (2003) Gene therapy for Parkinson's disease. J Neural Transm Suppl: 205–213

Morelli M (2003) Adenosine A2A antagonists: potential preventive and palliative treatment for Parkinson's disease. Exp Neurol 184:20–23

Muralikrishnan D, Mohanakumar KP (1998) Neuroprotection by bromocriptine against 1-methyl-4-phenyl-1,2,3,6-tetrahydropyridine-induced neurotoxicity in mice. FASEB J 12:905–912

Murphy RC, Messer A (2004) A single-chain Fv intrabody provides functional protection against the effects of mutant protein in an organotypic slice culture model of Huntington's disease. Brain Res Mol Brain Res 121:141–145

Nakao N, Yokote H, Nakai K, Itakura T (2000) Promotion of survival and regeneration of nigral dopamine neurons in a rat model of Parkinson's disease after implantation of embryonal carcinoma-derived neurons genetically engineered to produce glial cell line-derived neurotrophic factor. J Neurosurg 92:659–670

Nutt JG (2003) Long-term L-DOPA therapy: challenges to our understanding and for the care of people with Parkinson's disease. Exp Neurol 184:9–13

Ogawa N, Tanaka K, Asanuma M, Kawai M, Masumizu T, Kohno M, Mori A (1994) Bromocriptine protects mice against 6-hydroxydopamine and scavenges hydroxyl free radicals in vitro. Brain Res 657:207–213
Ogomori K, Kitamoto T, Tateishi J, Sato Y, Suetsugu M, Abe M (1989) Beta-protein amyloid is widely distributed in the central nervous system of patients with Alzheimer's disease. Am J Pathol 134:243–251
Olanow CW, Tatton WG (1999) Etiology and pathogenesis of Parkinson's disease. Annu Rev Neurosci 22:123–144
Olanow CW, Watts RL, Koller WC (2001) An algorithm (decision tree) for the management of Parkinson's disease (2001): treatment guidelines. Neurology 56: S1-S88
Opacka-Juffry J, Ashworth S, Hume SP, Martin D, Brooks DJ, Blunt SB (1995) GDNF protects against 6-OHDA nigrostriatal lesion: in vivo study with microdialysis and PET. Neuroreport 7:348–352
Oransky I (2003) Gene therapy trial for Parkinson's disease begins. Lancet 362:712
Parkinson J (1817) An essay on the shaking palsy. London
Perrin RJ, Woods WS, Clayton DF, George JM (2000) Interaction of human alpha-Synuclein and Parkinson's disease variants with phospholipids. Structural analysis using site-directed mutagenesis. J Biol Chem 275:34393–34398
Polymeropoulos MH (2000) Genetics of Parkinson's disease. Ann N Y Acad Sci 920:28–32
Polymeropoulos MH, Higgins JJ, Golbe LI, Johnson WG, Ide SE, Di Iorio G, Sanges G, Stenroos ES, Pho LT, Schaffer AA, Lazzarini AM, Nussbaum RL, Duvoisin RC (1996) Mapping of a gene for Parkinson's disease to chromosome 4q21-q23. Science 274:1197–1199
Polymeropoulos MH, Lavedan C, Leroy E, Ide SE, Dehejia A, Dutra A, Pike B, Root H, Rubenstein J, Boyer R, Stenroos ES, Chandrasekharappa S, Athanassiadou A, Papapetropoulos T, Johnson WG, Lazzarini AM, Duvoisin RC, Di Iorio G, Golbe LI, Nussbaum RL (1997) Mutation in the alpha-synuclein gene identified in families with Parkinson's disease. Science 276:2045–2047
Przedborski S, Jackson-Lewis V, Naini AB, Jakowec M, Petzinger G, Miller R, Akram M (2001) The parkinsonian toxin 1-methyl-4-phenyl-1,2,3,6-tetrahydropyridine (MPTP): a technical review of its utility and safety. J Neurochem 76:1265–1274
Rochet JC, Outeiro TF, Conway KA, Ding TT, Volles MJ, Lashuel HA, Bieganski RM, Lindquist SL, Lansbury PT (2004) Interactions among alpha-synuclein, dopamine, and biomembranes: some clues for understanding neurodegeneration in Parkinson's disease. J Mol Neurosci 23:23–34

Sariola H, Saarma M (2003) Novel functions and signalling pathways for GDNF. J Cell Sci 116:3855–3862

Schapira AH, Olanow CW (2003) Rationale for the use of dopamine agonists as neuroprotective agents in Parkinson's disease. Ann Neurol 53 Suppl 3:S149–157; discussion S157–159

Scherfler C, Khan NL, Pavese N, Eunson L, Graham E, Lees AJ, Quinn NP, Wood NW, Brooks DJ, Piccini PP (2004) Striatal and cortical pre- and postsynaptic dopaminergic dysfunction in sporadic parkin-linked parkinsonism. Brain 127(Pt 6):1332–1342

Sebastiao AM, Ribeiro JA (2000) Fine-tuning neuromodulation by adenosine. Trends Pharmacol Sci 21:341–346

Seidler A, Hellenbrand W, Robra BP, Vieregge P, Nischan P, Joerg J, Oertel WH, Ulm G, Schneider E (1996) Possible environmental, occupational, and other etiologic factors for Parkinson's disease: a case-control study in Germany. Neurology 46:1275–1284

Shults CW, Oakes D, Kieburtz K, Beal MF, Haas R, Plumb S, Juncos JL, Nutt J, Shoulson I, Carter J, Kompoliti K, Perlmutter JS, Reich S, Stern M, Watts RL, Kurlan R, Molho E, Harrison M, Lew M (2002) Effects of coenzyme Q_{10} in early Parkinson disease: evidence of slowing of the functional decline. Arch Neurol 59:1541–1550

Singleton A, Gwinn-Hardy K, Sharabi Y, Li ST, Holmes C, Dendi R, Hardy J, Crawley A, Goldstein DS (2004) Association between cardiac denervation and parkinsonism caused by alpha-synuclein gene triplication. Brain 127:768–772

Singleton AB, Farrer M, Johnson J, Singleton A, Hague S, Kachergus J, Hulihan M, Peuralinna T, Dutra A, Nussbaum R, Lincoln S, Crawley A, Hanson M, Maraganore D, Adler C, Cookson MR, Muenter M, Baptista M, Miller D, Blancato J, Hardy J, Gwinn-Hardy K (2003) alpha-Synuclein locus triplication causes Parkinson's disease. Science 302:841

Song DD, Shults CW, Sisk A, Rockenstein E, Masliah E (2004) Enhanced substantia nigra mitochondrial pathology in human alpha-synuclein transgenic mice after treatment with MPTP. Exp Neurol 186:158–172

Stocchi F (2003) Prevention and treatment of motor fluctuations. Parkinsonism Relat Disord 9 Suppl 2:S73–81

Suenaga T, Hirano A, Llena JF, Ksiezak-Reding H, Yen SH, Dickson DW (1990) Modified Bielschowsky and immunocytochemical studies on cerebellar plaques in Alzheimer's disease. J Neuropathol Exp Neurol 49:31–40

Svenningsson P, Nomikos GG, Ongini E, Fredholm BB (1997) Antagonism of adenosine A2A receptors underlies the behavioural activating effect of caffeine and is associated with reduced expression of messenger RNA for NGFI-A and NGFI-B in caudate-putamen and nucleus accumbens. Neuroscience 79:753–764

Takashima H, Tsujihata M, Kishikawa M, Freed WJ (1999) Bromocriptine protects dopaminergic neurons from levodopa-induced toxicity by stimulating D(2)receptors. Exp Neurol 159:98–104

Takenouchi T, Hashimoto M, Hsu LJ, Mackowski B, Rockenstein E, Mallory M, Masliah E (2001) Reduced neuritic outgrowth and cell adhesion in neuronal cells transfected with human alpha-synuclein. Mol Cell Neurosci 17:141–150

Tarnopolsky MA, Beal MF (2001) Potential for creatine and other therapies targeting cellular energy dysfunction in neurological disorders. Ann Neurol 49:561–574

Thiruchelvam M, Brockel BJ, Richfield EK, Baggs RB, Cory-Slechta DA (2000a) Potentiated and preferential effects of combined paraquat and maneb on nigrostriatal dopamine systems: environmental risk factors for Parkinson's disease? Brain Res 873:225–234

Thiruchelvam M, Richfield EK, Baggs RB, Tank AW, Cory-Slechta DA (2000b) The nigrostriatal dopaminergic system as a preferential target of repeated exposures to combined paraquat and maneb: implications for Parkinson's disease. J Neurosci 20:9207–9214

Thiruchelvam M, Richfield EK, Goodman BM, Baggs RB, Cory-Slechta DA (2002) Developmental exposure to the pesticides paraquat and maneb and the Parkinson's disease phenotype. Neurotoxicology 23:621–633

Tomac A, Lindqvist E, Lin LF, Ogren SO, Young D, Hoffer BJ, Olson L (1995) Protection and repair of the nigrostriatal dopaminergic system by GDNF in vivo. Nature 373:335–339

Vigna E, Cavalieri S, Ailles L, Geuna M, Loew R, Bujard H, Naldini L (2002) Robust and efficient regulation of transgene expression in vivo by improved tetracycline-dependent lentiviral vectors. Mol Ther 5:252–261

Vu TQ, Ling ZD, Ma SY, Robie HC, Tong CW, Chen EY, Lipton JW, Carvey PM (2000) Pramipexole attenuates the dopaminergic cell loss induced by intraventricular 6-hydroxydopamine. J Neural Transm 107:159–176

Warner TT, Schapira AH (2003) Genetic and environmental factors in the cause of Parkinson's disease. Ann Neurol 53 Suppl 3:S16–23; discussion S23–25

Weinreb PH, Zhen W, Poon AW, Conway KA, Lansbury, PT Jr. (1996) NACP, a protein implicated in Alzheimer's disease and learning, is natively unfolded. Biochemistry 35:13709–13815

Whone, AL, Moore RY, Piccini PP, Brooks DJ (2003) Plasticity of the nigropallidal pathway in Parkinson's disease. Ann Neurol 53:206–213

Zou L, Jankovic J, Rowe DB, Xie W, Appel SH, Le W (1999) Neuroprotection by pramipexole against dopamine- and levodopa-induced cytotoxicity. Life Sci 64:1275–1285

10 Measuring Injury and Repair of Myelin and Neurons in Multiple Sclerosis

D.L. Arnold, J. Chen

10.1 Introduction

Protection and repair in MS can be considered in terms of both neurons (soma, dendrites, and axons) and myelin (oligodendrocytes). Although the integrity of neurons and myelin are usually tightly linked, they can be dissociated to some extent, and it is useful to consider separately the ways that the integrity of each can be measured in vivo using magnetic resonance techniques.

10.2 Measuring Neuronal Integrity In Vivo

A number of techniques are often used to provide biomarkers of neuronal integrity. MR spectroscopic measurement of NAA offers the most specific biomarker for neurons, but suffers from poor resolution and low signal-to-noise ratio. Other biomarkers of neuronal integrity, such as brain volume and T1 "black hole" evolution are less specific but have greater resolution and, in the case of atrophy estimation, at least, greater precision.

10.3 Magnetic Resonance Spectroscopy

Pathological specificity for injury to neurons (including dendrites and axons) can be provided by quantification of the neuronal marker compound, N-acetylaspartate using proton magnetic resonance spectroscopy (MRS; Arnold et al. 1990). 1H-MRS is fundamentally different from water-proton-based MRI techniques in that it acquires signals from protons in metabolites that are present in tissue at concentrations more than one thousand times lower than that of tissue water. The signal-to-noise ratio and image resolution that is possible in these metabolite-based images is much lower than that for water-based images. However, the resulting images provide chemical-pathological specificity that is not possible with conventional MR images. The various approaches to in vivo 1H-MRS include: (a) single-voxel 1H-MRS studies (in which proton spectra are acquired from a single volume, which may be as small as about 1 cc or as large as the whole brain); and (b) spectroscopic imaging studies (in which proton spectra are obtained from multiple volume elements at the same time, generating metabolite maps).

10.4 Proton MRS Measurement of NAA

The proton MRS spectrum of the normal human brain that is recorded at relatively long echo times (usually 136–272 ms) reveals three major resonance peaks (the locations of which are expressed as the difference in parts per million (ppm) between the resonance frequency of the compound of interest and that of a standard compound). These peaks are commonly ascribed to the following metabolites: (a) tetramethyl amines (Cho), which resonate at 3.2 ppm and are mostly choline-containing phospholipids that participate in membrane synthesis and degradation; (b) creatine and phosphocreatine (Cr), which resonate at 3.0 ppm and play an important role in energy metabolism; and (c) N-acetyl groups (NA), which resonate at 2.0 ppm and are comprised primarily of the neuronally- localized compound N-acetylaspartate (NAA) in the brain. A fourth peak usually arising from the methyl resonance of lactate (LA) or lipids (because both resonate at 1.3 ppm), is normally only-barely visible above the baseline noise but can be detected in certain pathological conditions. Spectra acquired at shorter echo times (e.g., 20–30 ms) are better for detecting resonances that have a short T2, including lipids and inositol. They also record broad, overlapping signals that complicate quantification of resonance intensities.

The simplest approach to the quantitation of 1H-MR spectra is to normalize the NA and Cho signal intensities to the signal intensity from Cr in the same voxel .This method does not provide absolute quantification and the resulting measures of relative concentration are only valid if the underlying pathology does not substantially affect the local concentration of Cr. Thus, it is important that Cr concentrations are relatively constant throughout normal brain tissue and that they have also been shown to be relatively constant in both the chronic lesions and the normal-appearing brain tissue of patients with MS (Caramanos et al. 2003). It should be noted, however, that Cr values have been shown to decrease in acute (De Stefano et al. 1995) and severely-hypointense lesions (van Walderveen et al. 1999). Thus, it is inappropriate to normalize within-lesion NA and Cho values to within-lesion Cr values in either acute lesions or T1-weighted black holes.

10.5 Proton MRS of MS

The resonance intensity that is ascribed to NAA is important in the characterization of MS pathology because NAA is localized exclusively within neurons and neuronal processes, such as axons and dendrites (Moffett et al. 1991; Moffett and Namboodiri 1995) in the adult brain. The validity of NAA as an axon-specific biomarker in the adult CNS, even in the presence of the usual responses of oligodendroglial cells to injury, has been confirmed in a model of rat optic nerve transection (Bjartmar et al. 2002). The validity of NAA as a surrogate for axonal density in MS also has been confirmed in studies that have correlated (a) findings from in vivo MRS and histopathological analysis of cerebral biopsy specimens (Bitsch et al. 1999); and (b) findings from HPLC and histopathological analysis of spinal cord biopsy specimens (Bjartmar et al. 2000).

1H-MRSI-measured NA/Cr values have been used to quantify neuronal and axonal integrity in vivo in the brains of patients with MS starting with the observations of Arnold et al. (Arnold et al. 1990). MRS studies have shown that NA/Cr values are low in lesions and, to a lesser extent, in normal-appearing white matter (Arnold et al. 1992; Fu et al. 1998). Cerebral NAA is lower in patients with SPMS than patients with RRMS (Fu et al. 1998). Interestingly, this latter finding seems to be related more to differences in NAWM than in lesions. MRS- measured values of NA/Cr within the cortical NAGM of patients with MS have also been shown to be decreased relative to those in the cortical grey matter of healthy normal controls (Kapeller et al. 2001; Sharma et al. 2001; Adalsteinsson et al. 2003; Di Maio et al. 2003).

NA/Cr density decreases approximately 5 % per year in patients with RR MS (Fu et al. 1998) with a coefficient of variation of 5 %–10 %. In these patients, decreases in periventricular-NA/Cr values are strongly related to both their disease duration and their EDSS scores (De Stefano et al. 2001). The density of NA/Cr levels off in patients with SP MS (Fu et al. 1998), possibly because these patients develop atrophy in proportion to ongoing neuronal loss.

10.6 Brain Atrophy as a Biomarker of Neuronal Loss

Atrophy of the brain or spinal cord at postmortem examination is one of the pathological hallmarks of irreversible CNS damage and, as such, is related to loss of axons and myelin.

10.7 Relationship of Neuronal Integrity and Atrophy

Collins et al. (Collins et al. 2000) found that cerebral atrophy was correlated with axonal injury (as measured by decreases in 1H-MRSI NA/Cr values) in their group of patients with SPMS and that this relationship was different in RRMS. These findings suggest that decreases of neuronal integrity are related to decreases in brain volume. Importantly, however, there seems to be a decoupling between axonal injury and atrophy in the very early stages of the disease. It is not clear to what extent this decoupling is structural, related to physical decreases in the relative partial volume of neurons, and to what extent it is related to metabolic dysfunction of neurons early in the course of MS. This distinction could have important implications for strategies aimed at neuronal protection.

It should be noted that current MRI analysis techniques allow for the measurement of small changes in volume on the order of 0.2 % of total brain volume – changes of magnitude that are much smaller than those that can be identified on gross pathological examination. Thus far, it has been tempting to (a) assume that these small changes in brain volume have the same pathological significance as atrophy on postmortem examination and (b) suggest that they provide a measure of a specific pathological feature such as axonal loss. Unfortunately, it is not necessarily true that this is always the case: for example, myelin loss, glial- and matrix-related changes, as well as shifts in water distribution all occur in MS and may be associated with volume changes of this magnitude. While it is clear that volume measurements must contribute in some way to estimating the full extent of irreversible axonal damage in MS, further investigations are required to understand the precise pathological significance of atrophy and the mechanisms that contribute to its progression at different stages in the evolution of MS.

10.8 Measurement of Atrophy in Patients with MS

A variety of methods have been developed for measuring brain atrophy. For the purpose of assessing the effect of an intervention, methods designed to detect changes in volume with high precision are preferred. Techniques that measure the brain parenchymal fraction (BPF; Fisher et al. 1999) or the percent brain volume change (Siena; Smith et al. 2001) can measure change with a precision of 0.2 %–0.4 %.

10.9 Development of Atrophy in MS

The brains of MS patients atrophy by approximately 0.5 %–1 % per year (Rudick et al. 1999). This process begins at the onset of disease (Chard et al. 2002; Dalton et al. 2002) and may accelerate in the later stages of the disease (Collins et al. 2000). Importantly, the atrophy is not restricted to white matter, but involves the cortex, as well. Cortical atrophy may dominate in some circumstances (Davies et al. 2004; Dalton et al. 2004)

10.10 Diffusion-Weighted Imaging as a Biomarker of Axonal Integrity

It is possible to arrange for the signal intensity in MR images to be dependent on the diffusion of water: this technique is called diffusion weighted imaging (DWI; Le Bihan and Breton 1985). Depending on the sequence, the diffusion-weighted signal intensity can be made to reflect either the mean apparent diffusion coefficient (ADC) or the diffusivity in one or more specific directions. Appropriate postprocessing techniques, such as those used in diffusion tensor imaging (DTI; Basser et al. 1994), can produce maps of the preferred direction of diffusion and the degree to which diffusion is anisotropic. Diffusion anisotropy depends on restriction of diffusion in certain directions more than others owing to the presence of ordered, oriented structures. In white matter, this order is provided by neuronal fiber tracts. The extent to which restriction depends on the internal structure of axons, the axonal membrane or the myelin sheath is not entirely clear and may vary depending on the MR pulse sequences and gradient strengths used.

As current methods for obtaining images based on diffusion anisotropy have relatively low spatial resolution, crossing fibers within a voxel tend to confound the estimate. Thus, the use of diffusion anisotropy as a surrogate for axonal integrity is most reliable in larger fiber tracts that completely fill voxels, e.g., the corpus callosum or corticospinal tract. Sophisticated image processing can reconstruct fiber tracts in these images (Mori and van Zijl 2002). The image processing involved is complex and caution is indicated in the quantitative interpretation of such images.

10.11 Diffusion-Weighted Imaging in MS

DWI of MS lesions shows increases in diffusivity in NAWM and in lesions (Horsfield and Jones 2002). In some cases, the diffusivity in lesions approaches that of pure water. Histopathological correlations are not available. However, increases in diffusivity in lesions correlate with decreases in MTR and hypo-intensity of "black holes". Diffusion anisotropy is also reduced in MS lesions, possibly reflecting either disruption of the normal spatial order due to loss of axons, edema, or gliosis.

10.12 Measuring Myelin Integrity In Vivo

Remyelination not only improves conduction, but also is neuroprotective. Remyelinated lesions show less axonal injury that demyelinated lesions, (Kuhlmann et al. 2002) and demyelinated axons may be prone to early death (Bjartmar and Trapp 2001). Thus, it is appropriate to consider myelin integrity as an important aspect of neuroprotection.

10.13 "Myelin Water" as a Biomarker of Myelin Integrity

Two nonconventional MRI metrics are currently under active investigation of their ability to provide a biomarker of myelin content with acceptable pathological specificity. These are: (a) the short T2 component of water and (b) magnetization transfer.

The T2 relaxation time of water in tissue is not homogeneous. Bulk water such as CSF has a T2 on the order of 1 s. Water in the extracellular and intracellular space has a T2 on the order of 100 ms. Water trapped between the lamellae of myelin (and probably within myelin vesicles, as well) has a T2 on the order of 20 ms. The signal intensity on conventional T2-weighted MRI, normally obtained with echo times of 10–100 ms, is determined by a weighted average of T2 components that are present and have a relaxation times in this range. Since the water bound to myelin ("myelin water") makes up only about 10 % of tissue water in white matter, the T2-weighted signal is dominated by extracellular water and intracellular water, and is relatively insensitive to myelin water. It is possible, however, using customized acquisition techniques, to specifically quantify the component of water with T2 around 20 ms that putatively originates from water trapped in myelin.

MacKay et al. have pioneered T2 relaxometry studies in MS (MacKay et al. 1994; Moore et al. 2000).

10.14 Myelin Water Imaging in MS

Myelin water imaging has not been widely performed due to the long scan times required. The myelin water percentage is decreased in both NAWM and lesions of patients with MS. The relationship to MTR and myelin integrity versus myelin content is not straightforward and is still an area of active research.

10.15 MT Imaging in MS

Magnetization transfer (MT) is an MR phenomenon in which spins in two or more distinct environments exchange their magnetization via cross relaxation and/or chemical exchange (Edzes and Samulski 1978; Fung 1986; Grad et al. 1991).

MT imaging (Wolff and Balaban 1989), by virtue of the fact that it is based on the exchange with water bound to macromolecules, has specificity for myelin (Dousset et al. 1992, 1995; Brochet and Dousset 1999; Deloire-Grassin et al. 2000). The precise level of specificity will of course depend to some extent on the nature of the pathology being evaluated, and the way in which the MT contrast is generated and analyzed. The majority of human MT imaging studies have been performed using pulses designed to "saturate" the short T2, semisolid spins, which have a broad linewidth compared to bulk water. The MT effect can then be detected by observing the effects of the transfer of this saturation to observable, liquid spins. To isolate the effects of MT, MRI acquisitions are often performed with and without saturating pulses. Ratio or percent difference images (MTR images) are then calculated. These semiquantitative images reflect a complex combination of sequence and relaxation parameters in addition to pure MT (Pike et al. 1993; Pike 1996).

MT is well established as an important relaxation pathway in white matter(Dousset et al. 1992; Koenig et al. 1990; Beaulieu and Allen 1994; Fralix et al. 1991; Kucharczyk et al. 1994). Reductions in MTR are widely accepted as indicating demyelination and tissue damage (Brochet and Dousset 1999; Filippi 1999; Barkhof and van Walderveen 1999). This is supported by several lines of evidence, including: animal models with histopathological correlations (Dousset et al. 1992, 1995; Lexa et al. 1994), human studies of leukodystrophies (Silver et al. 1996), and postmortem studies of MS patients (Barkhof et al. 2003).

Acute inflammatory demyelination within MS lesions is associated with large reductions of MTR (Brochet and Dousset 1999; Filippi 1999). Diffuse, nonlesional decreases are also present throughout the NAWM. Focal decreases of MTR can be detected long before lesions appear on MRI (Pike et al. 2000).

Changes in MTR can be used to assess demyelination and remyelination of MS lesions in vivo. Methods for doing this in gadolinium-enhanced lesions, and in the same regions of brain before and after enhancement have been described by Richert et al. (Richert et al. 2001). Chen et al. (Chen et al. 2004) developed a voxel-based analysis of change in MTR in T2-weighted lesions in order to quantify separately the percent of T2-weighted lesion volume with ongoing demyelination and the percent of T2-weighted lesion volume with ongoing remyelination (Fig. 1). Although changes in magnetization transfer can be quantified absolutely using a series of customized acquisitions, acquisition of MTR images is straightforward and feasible on most current clinical MRI scanners.

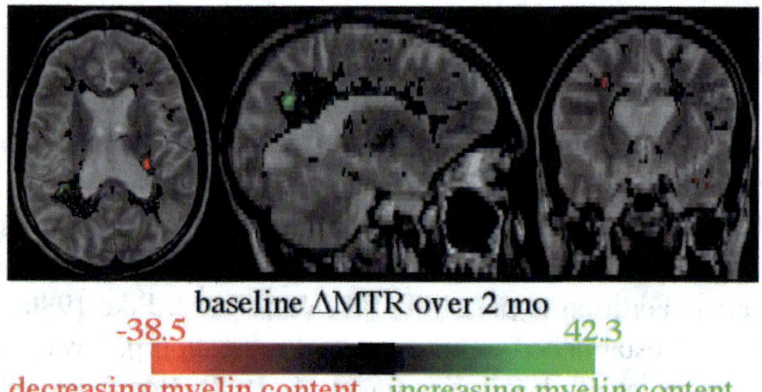

Fig. 1. Remyelinating lesion voxels (*green*) and demyelinating lesion voxels (*red*) determined by MTR imaging

10.16 Summary

Neuroprotection in MS needs to be considered in the context of several pathological processes: limitation of acute inflammatory injury to myelin and axons, remyelination, survival of demyelinated axons, and limitation of more diffuse, nonlesional pathology that affects myelin and axons. Advanced MRI techniques are capable of reporting on all of these different pathological features of MS and will be an important aspect of the assessment of neuroprotection strategies in MS, when these become available.

References

Adalsteinsson E, Langer-Gould A, Homer RJ, Rao A, Sullivan EV, Lima CA, et al. (2003) Gray matter N-acetyl aspartate deficits in secondary progressive but not relapsing-remitting multiple sclerosis. Am J Neuroradiol 24(10):1941–1945

Arnold DL, Matthews PM, Francis G, Antel J (1990) Proton magnetic resonance spectroscopy of human brain in vivo in the evaluation of multiple sclerosis: assessment of the load of disease. Magn Reson Med 14:154–159

Arnold DL, Matthews PM, Francis GS, O'Connor J, Antel JP (1992) Proton magnetic resonance spectroscopic imaging for metabolic characterization of demyelinating plaques. Ann Neurol 31:235–241

Barkhof F, van Walderveen M (1999) Characterization of tissue damage in multiple sclerosis by nuclear magnetic resonance. 354(1390):1675–1686

Barkhof F, Bruck W, De Groot CJ, Bergers E, Hulshof S, Geurts J, et al. (2003) Remyelinated lesions in multiple sclerosis: magnetic resonance image appearance. Arch Neurol 60(8):1073–1081

Basser PJ, Mattiello J, LeBihan D (1994) Estimation of the effective self-diffusion tensor from the NMR spin echo. J Magn Reson B 103(3):247–254

Beaulieu C, Allen PS (1994) Some magnetization transfer properties of water in myelinated and non-myelinated nerves. Proceedings of the Society of Magnetic Resonance in Medicine, vol 1. Wiley, New York, p 169

Bitsch A, Bruhn H, Vougioukas V, Stringaris A, Lassmann H, Frahm J, et al. (1999) Inflammatory CNS demyelination: histopathologic correlation with in vivo quantitative proton MR spectroscopy. AJNR Am J Neuroradiol 20(9):1619–1627

Bjartmar C, Trapp BD (2001) Axonal and neuronal degeneration in multiple sclerosis: mechanisms and functional consequences. Curr Opin Neurol 14(3):271–278

Bjartmar C, Kidd G, Mork S, Rudick R, Trapp BD (2000) Neurological disability correlates with spinal cord axonal loss and reduced N- acetyl aspartate in chronic multiple sclerosis patients. Ann Neurol 48(6):893–901

Bjartmar C, Battistuta J, Terada N, Dupree E, Trapp BD (2002) N- acetylaspartate is an axon-specific marker of mature white matter in vivo: a biochemical and immunohistochemical study on the rat optic nerve. Ann Neurol 51(1):51–58

Brochet B, Dousset V (1999) Pathological correlates of magnetization transfer imaging abnormalities in animal models and humans with multiple sclerosis. Neurology 53(5 Suppl 3):S12–S17

Caramanos Z, Narayanan S, Hum S, Arnold DL (2003) 1H-MRS quantitation of NA and Cr in the normal appearing and the lesional white matter of patients with MS. In: Proceedings of the International Society for Magnetic Resonance in Medicine, Eleventh Scientific Meeting, Toronto, Canada. p 273

Chard DT, Griffin CM, Parker GJ, Kapoor R, Thompson AJ, Miller DH (2002) Brain atrophy in clinically early relapsing-remitting multiple sclerosis. Brain 125(Pt 2):327–337

Chen JT, Arnold DL, Freedman MS, Atkins HL, Group CMBS (2004) The Impact of immunoablation with autologous stem cell transplant on lesion remyelination and demyelination in aggressive multiple sclerosis. In: Neurology; 62(7, Suppl 5):A351

Collins LD, Narayanan S, Caramanos Z, De Stefano N, Tartaglia MC, Arnold DL (2000) Relation of cerebral atrophy in multiple sclerosis to severity of disease and axonal injury [abstract]. Neurology 54(Suppl. 3):A17

Dalton CM, Brex PA, Jenkins R, Fox NC, Miszkiel KA, Crum WR, et al. (2002) Progressive ventricular enlargement in patients with clinically isolated syndromes is associated with the early development of multiple sclerosis. J Neurol Neurosurg Psychiatry 73(2):141–147

Dalton CM, Chard DT, Davies GR, Miszkiel KA, Altmann DR, Fernando K, et al. (2004) Early development of multiple sclerosis is associated with progressive grey matter atrophy in patients presenting with clinically isolated syndromes. Brain 127(Pt 5):1101–1107

Davies GR, Ramio-Torrenta L, Hadjiprocopis A, Chard DT, Griffin CM, Rashid W, et al. (2004) Evidence for grey matter MTR abnormality in minimally disabled patients with early relapsing-remitting multiple sclerosis. J Neurol Neurosurg Psychiatry 75(7):998–1002

De Stefano N, Matthews PM, Antel JP, Francis G, Arnold DL (1995) Proton MR spectroscopic imaging of acute demyelinating lesions: Chemical- pathological changes and their correlation with clinical disability. Ital J Neurol Sci (Supp. 7):46

De Stefano N, Narayanan S, Francis GS, Arnaoutelis R, Tartaglia MC, Antel JP, et al. (2001) Evidence of axonal damage in the early stages of multiple sclerosis and its relevance to disability. Arch Neurol 58:65–70

Deloire-Grassin MS, Brochet B, Quesson B, Delalande C, Dousset V, Canioni P, et al. (2000) In vivo evaluation of remyelination in rat brain by magnetization transfer imaging. J Neurol Sci 178(1):10–16

Di Maio S, Narayanan S, De Stefano N, chen J, Lapierre Y, Arnold DL (2003) In vivo evidence of cortical neuronal injury in secondary progressive multiple sclerosis. In: International Society for Magnetic Resonance in Medicine; 2003 May 11, Toronto, p. 1998

Dousset V, Grossman RI, Ramer KN, Schnall MD, Young LH, Gonzalez-Scarano F, et al. (1992) Experimental allergic encephalomyelitis and multiple sclerosis: lesion characterization with magnetization transfer imaging. Radiology 182(2):483–491

Dousset V, Brochet B, Vital A, Gross C, Benazzouz A, Boullerne A, et al. (1995) Lysolecithin-induced demyelination in primates: preliminary in vivo study with MR and magnetization transfer. AJNR Am J Neuroradiol 16(2):225–231

Edzes HT, Samulski ED (1978) The measurement of cross-relaxation effects in the proton NMR spin-lattice relaxation of water in biological systems: Hydrated collagen and muscle. J Magn Res 31(2):207–229

Filippi M (1999) Magnetization transfer imaging to monitor the evolution of individual multiple sclerosis lesions. Neurology 53(5 Suppl 3):S18–S22

Fisher E, Rudick RA, Tkach JA, Lee JC, Masaryk TJ, Simon JH, et al. (1999) Automated calculation of whole brain atrophy from magnetic resonance images for monitoring multiple sclerosis. Neurology 52(6):A352–A353

Fralix TA, Ceckler TL, Wolff SD, Simon SA, Balaban RS (1991) Lipid bilayer and water proton magnetization transfer: effect of cholesterol. Magn Res Med 18(1):214–223

Fu L, Matthews PM, De Stefano N, Worsley KJ, Narayanan S, Francis GS, et al. (1998) Imaging axonal damage of normal appearing white matter in multiple sclerosis. Brain 121:103–113

Fung B (1986) Nuclear magnetic resonance study of water interactions with proteins, biomolecules, membranes and tissues. In: Packer L (ed) Methods in enzymology. Academic, San Diego, p 151

Grad J, Mendelson D, Hyder F, Bryant RG (1991) Applications of nuclear magnetic cross-relaxation spectroscopy to tissues. Magn Res Med 17:452–459

Horsfield MA, Jones DK (2002) Applications of diffusion-weighted and diffusion tensor MRI to white matter diseases—a review. NMR Biomed 15(7–8):570–577

Kapeller P, McLean MA, Griffin CM, Chard D, Parker GJ, Barker GJ, et al. (2001) Preliminary evidence for neuronal damage in cortical grey matter and normal appearing white matter in short duration relapsing-remitting multiple sclerosis: a quantitative MR spectroscopic imaging study. J Neurol 248(2):131–138

Koenig SH, Brown RD, III, Spiller M, Lundbom N (1990) Relaxometry of brain: why white matter appears bright in MRI. 14(3):482–495

Kucharczyk W, Macdonald PM, Stanisz GJ, Henkelman RM (1994) Relaxivity and magnetization transfer of white matter lipids at MR imaging: importance of cerebrosides and pH. Radiology 192(2):521–529

Kuhlmann T, Lingfeld G, Bitsch A, Schuchardt J, Bruck W (2002) Acute axonal damage in multiple sclerosis is most extensive in early disease stages and decreases over time. Brain 125(Pt 10):2202–2212

Le Bihan D, Breton E (1985) Diffusion imaging in vivo by nuclear magnetic resonance. CR Acad Sci Paris 301:1109–1112

Lexa FJ, Grossman RI, Rosenquist AC (1994) Dyke Award paper. MR of wallerian degeneration in the feline visual system: characterization by magnetization transfer rate with histopathologic correlation. AJNR. Am J Neuroradiol 15(2):201–212

MacKay A, Whittall K, Adler J, Li D, Paty D, Graeb D (1994) In vivo visualization of myelin water in brain by magnetic resonance. 31(6):673–677

Moffett JR, Namboodiri MA (1995) Differential distribution of N- acetylaspartylglutamate and N-acetylaspartate immunoreactivities in rat forebrain. J Neurocytol 24(6):409–433

Moffett JR, Namboodiri MA, Cangro CB, Neale JH (1991) Immunohistochemical localization of N-acetylaspartate in rat brain. Neuroreport 2(3):131–134

Moore GR, Leung E, MacKay AL, Vavasour IM, Whittall KP, Cover KS, et al. (2000) A pathology-MRI study of the short-T2 component in formalin-fixed multiple sclerosis brain. Neurology 55(10):1506–1510

Mori S, van Zijl PC (2002) Fiber tracking: principles and strategies—a technical review. NMR Biomed 15(7–8):468–480

Pike GB (1996) Pulsed magnetization transfer contrast in gradient echo imaging: a two-pool analytic description of signal response. Magn Reson Med 36(1):95–103

Pike GB, Glover GH, Hu BS, Enzmann DR (1993) Pulsed magnetization transfer spin-echo MR imaging. 3(3):531–539

Pike GB, De Stefano N, Narayanan S, Worsley KJ, Pelletier D, Francis GS, et al. (2000) Multiple sclerosis: magnetization transfer MR imaging of white matter before lesion appearance on T2-weighted images. Radiology 215(3):824–830

Richert ND, Ostuni JL, Bash CN, Leist TP, McFarland HF, Frank JA (2001) Interferon beta-1b and intravenous methylprednisolone promote lesion recovery in multiple sclerosis. Mult Scler 7(1):49–58

Rudick RA, Fisher E, Lee JC, Simon J, Jacobs L (1999) Use of the brain parenchymal fraction to measure whole brain atrophy in relapsing-remitting MS. Multiple Sclerosis Collaborative Research Group. Neurology 53(8):1698–1704

Sharma R, Narayana PA, Wolinsky JS (2001) Grey matter abnormalities in multiple sclerosis: proton magnetic resonance spectroscopic imaging. Mult Scler 7(4):221–226

Silver NC, Barker GJ, MacManus DG, Miller DH, Thorpe JW, Howard RS (1996) Decreased magnetisation transfer ratio due to demyelination: a case of central pontine myelinolysis. J Neurol Neurosurg Psychiatry 61(2):208–209

Smith S, Zhang YY, Jenkinson M, Matthews PM, De Stefano N (2001) SIENA: Single and multiple time point brain atrophy analysis. NeuroImage 13(6):S250

van Walderveen MA, Barkhof F, Pouwels PJ, van Schijndel RA, Polman CH, Castelijns JA (1999) Neuronal damage in T1-hypointense multiple sclerosis lesions demonstrated in vivo using proton magnetic resonance spectroscopy. Ann Neurol 46(1):79–87

Wolff SD, Balaban RS (1989) Magnetization transfer contrast (MTC) and tissue water proton relaxation in vivo. Magn Res Med 10:135–144

Ernst Schering Research Foundation Workshop

Editors: Günter Stock
Monika Lessl

Vol. 1 (1991): Bioscience ⇌ Societly Workshop Report
Editors: D.J. Roy, B.E. Wynne, R.W. Old

Vol. 2 (1991): Round Table Discussion on Bioscience ⇌ Society
Editor: J.J. Cherfas

Vol. 3 (1991): Excitatory Amino Acids and Second Messenger Systems
Editors: V.I. Teichberg, L. Turski

Vol. 4 (1992): Spermatogenesis – Fertilization – Contraception
Editors: E. Nieschlag, U.-F. Habenicht

Vol. 5 (1992): Sex Steroids and the Cardiovascular System
Editors: P. Ramwell, G. Rubanyi, E. Schillinger

Vol. 6 (1993): Transgenic Animals as Model Systems for Human Diseases
Editors: E.F. Wagner, F. Theuring

Vol. 7 (1993): Basic Mechanisms Controlling Term and Preterm Birth
Editors: K. Chwalisz, R.E. Garfield

Vol. 8 (1994): Health Care 2010
Editors: C. Bezold, K. Knabner

Vol. 9 (1994): Sex Steroids and Bone
Editors: R. Ziegler, J. Pfeilschifter, M. Bräutigam

Vol. 10 (1994): Nongenotoxic Carcinogenesis
Editors: A. Cockburn, L. Smith

Vol. 11 (1994): Cell Culture in Pharmaceutical Research
Editors: N.E. Fusenig, H. Graf

Vol. 12 (1994): Interactions Between Adjuvants, Agrochemical and Target Organisms
Editors: P.J. Holloway, R.T. Rees, D. Stock

Vol. 13 (1994): Assessment of the Use of Single Cytochrome P450 Enzymes in Drug Research
Editors: M.R. Waterman, M. Hildebrand

Vol. 14 (1995): Apoptosis in Hormone-Dependent Cancers
Editors: M. Tenniswood, H. Michna

Vol. 15 (1995): Computer Aided Drug Design in Industrial Research
Editors: E.C. Herrmann, R. Franke

Vol. 16 (1995): Organ-Selective Actions of Steroid Hormones
Editors: D.T. Baird, G. Schütz, R. Krattenmacher

Vol. 17 (1996): Alzheimer's Disease
Editors: J.D. Turner, K. Beyreuther, F. Theuring

Vol. 18 (1997): The Endometrium as a Target for Contraception
Editors: H.M. Beier, M.J.K. Harper, K. Chwalisz

Vol. 19 (1997): EGF Receptor in Tumor Growth and Progression
Editors: R.B. Lichtner, R.N. Harkins

Vol. 20 (1997): Cellular Therapy
Editors: H. Wekerle, H. Graf, J.D. Turner

Vol. 21 (1997): Nitric Oxide, Cytochromes P 450, and Sexual Steroid Hormones
Editors: J.R. Lancaster, J.F. Parkinson

Vol. 22 (1997): Impact of Molecular Biology and New Technical Developments in Diagnostic Imaging
Editors: W. Semmler, M. Schwaiger

Vol. 23 (1998): Excitatory Amino Acids
Editors: P.H. Seeburg, I. Bresink, L. Turski

Vol. 24 (1998): Molecular Basis of Sex Hormone Receptor Function
Editors: H. Gronemeyer, U. Fuhrmann, K. Parczyk

Vol. 25 (1998): Novel Approaches to Treatment of Osteoporosis
Editors: R.G.G. Russell, T.M. Skerry, U. Kollenkirchen

Vol. 26 (1998): Recent Trends in Molecular Recognition
Editors: F. Diederich, H. Künzer

Vol. 27 (1998): Gene Therapy
Editors: R.E. Sobol, K.J. Scanlon, E. Nestaas, T. Strohmeyer

Vol. 28 (1999): Therapeutic Angiogenesis
Editors: J.A. Dormandy, W.P. Dole, G.M. Rubanyi

Vol. 29 (2000): Of Fish, Fly, Worm and Man
Editors: C. Nüsslein-Volhard, J. Krätzschmar

Vol. 30 (2000): Therapeutic Vaccination Therapy
Editors: P. Walden, W. Sterry, H. Hennekes

Vol. 31 (2000): Advances in Eicosanoid Research
Editors: C.N. Serhan, H.D. Perez

Vol. 32 (2000): The Role of Natural Products in Drug Discovery
Editors: J. Mulzer, R. Bohlmann

Vol. 33 (2001): Stem Cells from Cord Blood, In Utero Stem Cell Development, and Transplantation-Inclusive Gene Therapy
Editors: W. Holzgreve, M. Lessl

Vol. 34 (2001): Data Mining in Structural Biology
Editors: I. Schlichting, U. Egner

Vol. 35 (2002): Stem Cell Transplantation and Tissue Engineering
Editors: A. Haverich, H. Graf

Vol. 36 (2002): The Human Genome
Editors: A. Rosenthal, L. Vakalopoulou

Vol. 37 (2002): Pharmacokinetic Challenges in Drug Discovery
Editors: O. Pelkonen, A. Baumann, A. Reichel

Vol. 38 (2002): Bioinformatics and Genome Analysis
Editors: H.-W. Mewes, B. Weiss, H. Seidel

Vol. 39 (2002): Neuroinflammation – From Bench to Bedside
Editors: H. Kettenmann, G.A. Burton, U. Moenning

Vol. 40 (2002): Recent Advances in Glucocorticoid Receptor Action
Editors: A. Cato, H. Schaecke, K. Asadullah

Vol. 41 (2002): The Future of the Oocyte
Editors: J. Eppig, C. Hegele-Hartung

Vol. 42 (2003): Small Molecule-Protein Interaction
Editors: H. Waldmann, M. Koppitz

Vol. 43 (2003): Human Gene Therapy: Present Opportunities and Future Trends
Editors: G.M. Rubanyi, S. Ylä-Herttuala

Vol. 44 (2004): Leucocyte Trafficking: The Role of Fucosyltransferases and Selectins
Editors: A. Hamann, K. Asadullah, A. Schottelius

Vol. 45 (2004): Chemokine Roles in Immunoregulation and Disease
Editors: P.M. Murphy, R. Horuk

Vol. 46 (2004): New Molecular Mechanisms of Estrogen Action and Their Impact on Future Perspectives in Estrogen Therapy
Editors: K.S. Korach, A. Hillisch, K.H. Fritzemeier

Vol. 47 (2004): Neuroinflammation in Stroke
Editors: U. Dirnagl, B. Elger

Vol. 48 (2004): From Morphological Imaging to Molecular Targeting
Editors: M. Schwaiger, L. Dinkelborg, H. Schweinfurth

Vol. 49 (2004): Molecular Imaging
Editors: A.A. Bogdanov, K. Licha

Vol. 50 (2005): Animal Models of T Cell-Mediated Skin Diseases
Editors: T. Zollner, H. Renz, K. Asadullah

Vol. 51 (2005): Biocombinatorial Approaches for Drug Finding
Editors: W. Wohlleben, T. Spellig, B. Müller-Tiemann

Vol. 52 (2005): New Mechanisms for Tissue-Selective Estrogen-Free Contraception
Editors: H.B. Croxatto, R. Schürmann, U. Fuhrmann, I. Schellschmidt

Vol. 53 (2005): Opportunities and Challanges of the Therapies Targeting CNS Regeneration
Editors: D. Perez, B. Mitrovic, A. Baron Van Evercooren

9783540240822